HARCOURT

Science

Harcourt School Publishers

Orlando • Boston • Dallas • Chicago • San Diego

www.harcourtschool.com

The **blue and yellow macaw** *(Ara ararauna)* lives in the trees of the rain forest in South America and Central America. It can grow to be about 33 in. (84 cm) in length. It is the largest member of the parrot family. Its favorite food is the seed of the fruit of one rain forest tree. Blue and yellow macaws often gather at "lick" areas to eat mineral- and salt-bearing clay. The inside covers of this book show a closeup of blue and yellow macaw feathers.

Printed in the United States of America

ISBN 0-15-325383-5 UNIT A
ISBN 0-15-325384-3 UNIT B
ISBN 0-15-325385-1 UNIT C
ISBN 0-15-325386-X UNIT D
ISBN 0-15-325387-8 UNIT E
ISBN 0-15-325388-6 UNIT F

3 4 5 6 7 8 9 10 032 10 09 08 07 06 05 04 03 02

Authors

Marjorie Slavick Frank
Former Adjunct Faculty Member
Hunter, Brooklyn, and Manhattan Colleges
New York, New York

Robert M. Jones
Professor of Education
University of Houston–Clear Lake
Houston, Texas

Gerald H. Krockover
Professor of Earth and Atmospheric Science Education
School Mathematics and Science Center
Purdue University
West Lafayette, Indiana

Mozell P. Lang
Science Education Consultant
Michigan Department of Education
Lansing, Michigan

Joyce C. McLeod
Visiting Professor
Rollins College
Winter Park, Florida

Carol J. Valenta
Vice President—Education, Exhibits, and Programs
St. Louis Science Center
St. Louis, Missouri

Barry A. Van Deman
Program Director, Informal Science Education
Arlington, Virginia

Senior Editorial Advisor

Napoleon Adebola Bryant, Jr.
Professor Emeritus of Education
Xavier University
Cincinnati, Ohio

Program Advisors

Michael J. Bell
Assistant Professor of Early Childhood Education
School of Education
West Chester University
West Chester, Pennsylvania

George W. Bright
Professor of Mathematics Education
The University of North Carolina at Greensboro
Greensboro, North Carolina

Pansy Cowder
Science Specialist
Tampa, Florida

Robert H. Fronk
Head, Science/Mathematics Education Department
Florida Institute of Technology
Melbourne, Florida

Bernard A. Harris, Jr.
Physician and Former Astronaut
(STS 55—Space Shuttle *Columbia*, STS 63—Space Shuttle *Discovery*)
President, The Harris Foundation
Houston, Texas

Lois Harrison-Jones
Education and Management Consultant
Dallas, Texas

Kenneth R. Mechling
Professor of Biology and Science Education
Clarion University of Pennsylvania
Clarion, Pennsylvania

Nancy Roser
Professor of Language and Literacy Studies
University of Texas, Austin
Austin, Texas

Program Advisor and Activities Writer

Barbara ten Brink
Science Director
Round Rock Independent School District
Round Rock, Texas

Reviewers and Contributors

Sidney Jay Abramowitz
District Administrator for Mathematics, Science & Technology
Stamford Public Schools
Stamford, Connecticut

Patricia L. Bush
Teacher
Jesse Stuart Elementary School
Madisonville, Kentucky

Kathryn Henry
Teacher Trainer
Public School CS 200
New York, New York

Jacqueline Howard
Teacher
Millard Hensley Elementary School
Salyersville, Kentucky

Faye Kelley
Teacher
Gardendale Elementary School
Gardendale, Alabama

Faye McCollum
Instructional Specialist
Muscogee County School District
Columbus, Georgia

Margie H. McCoy
Teacher
Salyer Elementary School
Salyersville, Kentucky

Chester H. Melcher
Supervisor of Curriculum and Instruction—Science
Racine Unified School District
Racine, Wisconsin

Michael F. Ryan
Educational Technology Specialist
Lake County Schools
Tavares, Florida

Michelle Bonk Severtson
Teacher
Olson Elementary
Bloomington, Minnesota

Stanley J. Wegrzynowski
Director of Science
Buffalo Public Schools
Buffalo, New York

Lee Ann White
Teacher
Morgantown Elementary School
Morgantown, Kentucky

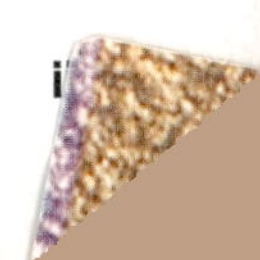

UNIT A

LIFE SCIENCE

A World of Living Things

UNIT B

LIFE SCIENCE

Looking at Ecosystems

UNIT C

EARTH SCIENCE

Earth's Surface

UNIT D

EARTH SCIENCE

Patterns on Earth and in Space

UNIT E

PHYSICAL SCIENCE
Matter and Energy

UNIT F

PHYSICAL SCIENCE

Forces and Motion

Planning an Investigation

How do scientists answer a question or solve a problem they have identified? They use organized ways called **scientific methods** to plan and conduct a study. They use science process skills to help them gather, organize, analyze, and present their information.

Nathan is using this scientific method for experimenting to find an answer to his question. You can use these steps, too.

STEP 1 Observe, and ask questions.

- Use your senses to make observations.
- Record **one** question that you would like to answer.
- Write down what you already know about the topic of your question.
- Decide what other information you need.
- Do research to find more information about your topic.

What soil works best for planting marigold seeds?
I need to find out more about the different kinds of soils.

STEP 2 Form a hypothesis.

- Write a possible answer, or hypothesis, to your question. A **hypothesis** is a possible answer that can be tested.
- Write your hypothesis in a complete sentence.

My hypothesis is: Marigold seeds sprout best in potting soil.

STEP 3 Plan an experiment.

- Decide how to conduct a fair test of your hypothesis by controlling variables. **Variables** are factors that can affect the outcome of the investigation.
- Write down the steps you will follow to do your test.
- List the equipment you will need.
- Decide how you will gather and record your data.

I'll put identical seeds in three different kinds of soil. Each flowerpot will get the same amount of water and light. So, I'll be controlling the variables of water and light.

STEP 4 Conduct the experiment.

- Follow the steps you wrote.
- Observe and measure carefully.
- Record everything that happens.
- Organize your data so you can study it carefully.

I'll measure each plant every 3 days. I'll record the results in a table and then make a bar graph to show the height of each plant 21 days after I planted the seeds.

STEP 5 Draw conclusions and communicate results.

- Analyze the data you gathered.
- Make charts, tables, or graphs to show your data.
- Write a conclusion. Describe the evidence you used to determine whether your test supported your hypothesis.
- Decide whether your hypothesis was correct.

Hmmm. My hypothesis was not correct. The seeds sprouted equally well in potting soil and sandy soil. They did not sprout at all in clay soil.

INVESTIGATE FURTHER

If your hypothesis was correct . . .

You may want to pose another question about your topic that you can test.

If your hypothesis was incorrect . . .

You may want to form another hypothesis and do a test of a different variable.

I'll test this new hypothesis: Marigold seeds sprout best in a combination of clay, sandy, and potting soil. I will plan and conduct a test using potting soil, sandy soil, and a combination of clay, sandy, and potting soil.

Do you think Nathan's new hypothesis is correct? Plan and conduct a test to find out!

Using Science Process Skills

When scientists try to find an answer to a question or do an experiment, they use thinking tools called **process skills.** You use many of the process skills whenever you speak, listen, read, write, or think. Think about how these students use process skills to help them answer questions, do experiments, and investigate the world around them.

What Sarah plans to investigate

Sarah collects seashells on her visit to the beach. She wants to make collections of shells that are alike in some way. She looks for shells of different sizes and shapes.

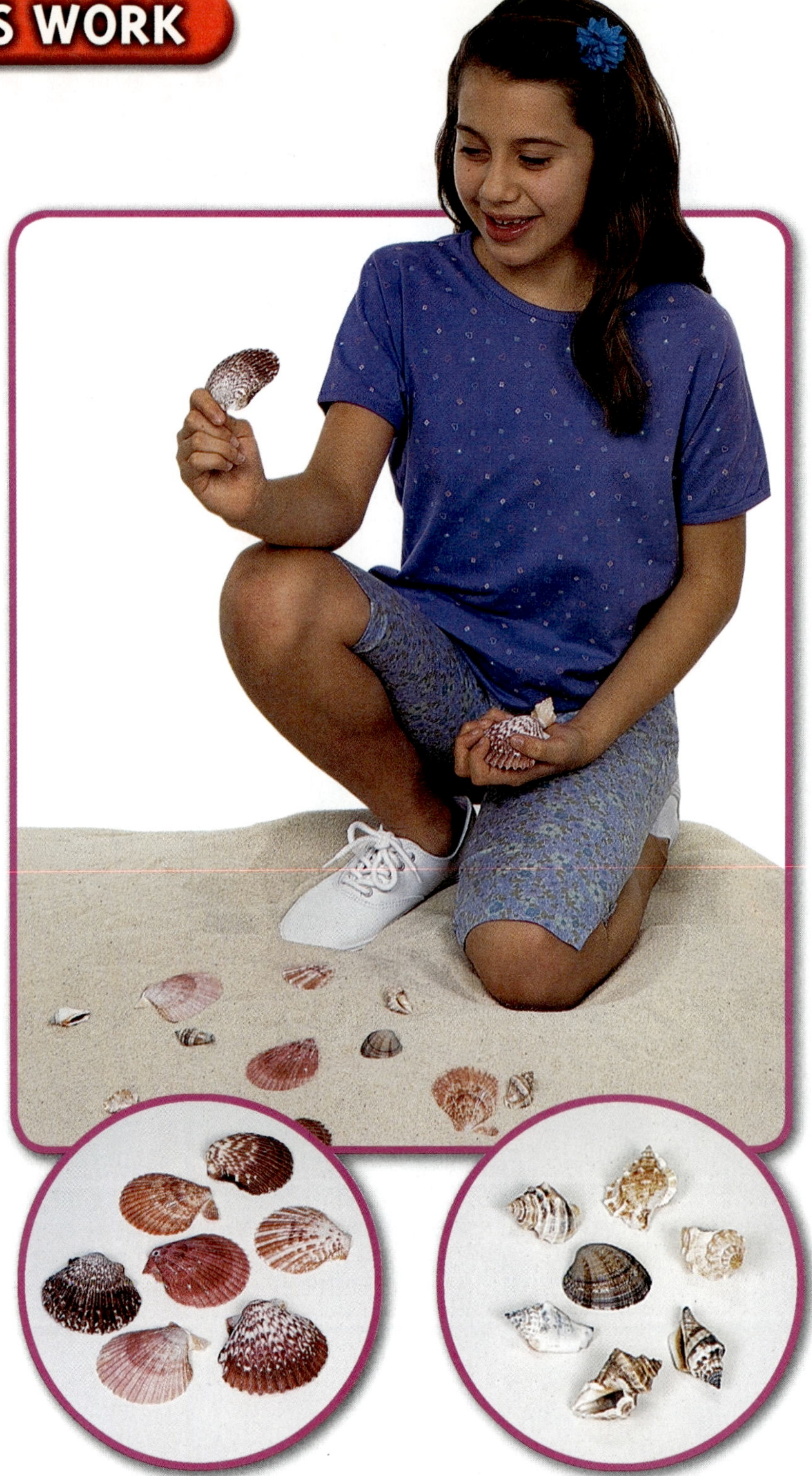

Process Skills

Observe—use the senses to learn about objects and events.

Compare—identify characteristics of things or events to find out how they are alike and different.

Classify—group or organize objects or events in categories based on specific characteristics.

How Sarah uses process skills

She **observes** the shells and **compares** their sizes, shapes, and colors. She **classifies** the shells first into groups based on their sizes and then into groups based on their shapes.

What Ling plans to investigate

Ling is interested in learning what makes the size and shape of a rock change. He plans an experiment to find out whether sand rubbing against a rock will cause pieces of the rock to flake off and change the size or shape of the rock.

How Ling uses process skills

He collects three rocks, **measures** their masses, and puts the rocks in a jar with sand and water. He shakes the rocks every day for a week. Then he measures and **records** the mass of the rocks, the sand, and the container. He interprets his data and concludes that rocks are broken down when sand rubs against them.

Process Skills

Measure—Compare an attribute of an object, such as mass, length, or capacity, to a unit of measure, such as gram, centimeter, or liter.

Gather, Record, Display, or Interpret Data

- Gather data by making observations that will be useful for inferences or predictions.
- Record data by writing down the observations in a table, graph, or notebook.
- Display data by making tables, charts, or graphs.
- Interpret data by drawing conclusions about what the data shows.

Process Skills

Use a Model—make a representation to help you understand an idea, an object, or an event, such as how something works.

Predict—form an idea of an expected outcome, based on observations or experience.

Infer—use logical reasoning to explain events and draw conclusions based on observations.

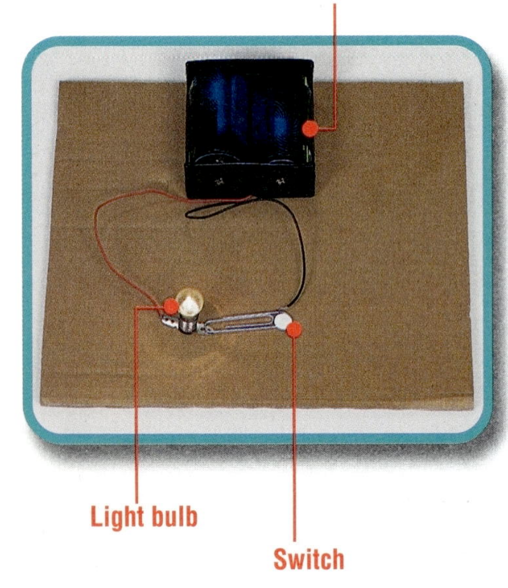

What Justin plans to investigate

Justin wants to find out how the light switch in his bedroom works. He uses batteries, a flashlight bulb, a bulb holder, thumbtacks, and a paper clip to help him.

How Justin uses process skills

He decides to **use a model** of the switch and the wires in the wall. He **predicts** that the bulb, wires, and batteries have to be connected to make the bulb light. He **infers** that moving the paper clip interrupts the flow of electricity and turns off the light. Justin's model verifies his prediction and inference.

What Kendra plans to investigate

Kendra wants to know what brand of paper towel absorbs the most water. She plans a test to find out how much water different brands of paper towels absorb. She can then tell her father which brand is the best one to buy.

How Kendra uses process skills

She chooses three brands of paper towels. She **hypothesizes** that one brand will absorb more water than the others. She **plans and conducts an experiment** to test her hypothesis, using the following steps:

- Pour 1 liter of water into each of three beakers.
- Put a towel from each of the three brands into a different beaker for 10 seconds.
- Pull the towel out of the water, and let it drain back into the beaker for 5 seconds.
- Measure the amount of water left in each beaker.

Kendra **controls variables** by making sure each beaker contains exactly the same amount of water and by timing each step in her experiment exactly.

Process Skills

Hypothesize—make a statement about an expected outcome.

Plan and Conduct an Experiment—identify and perform the steps necessary to test a hypothesis, using appropriate tools and recording and analyzing the data collected.

Control Variables—identify and control factors that affect the outcome of an experiment so that only one variable is changed in a test.

Reading to Learn

Scientists use reading, writing, and numbers in their work. They read to find out everything they can about a topic they are investigating. So it is important that scientists know the meanings of science vocabulary and that they understand what they read. Use the following strategies to help you become a good science reader!

Before Reading

- Read the **Find Out** statement to help you know what to look for as you read.
- Think: I need to find out what the parts of an ecosystem are and how they are organized.

- Look at the **Vocabulary** words.
- Be sure that you can pronounce each word.
- Look up each word in the Glossary.
- Say the definition to yourself. Use the word in a sentence to show its meaning.

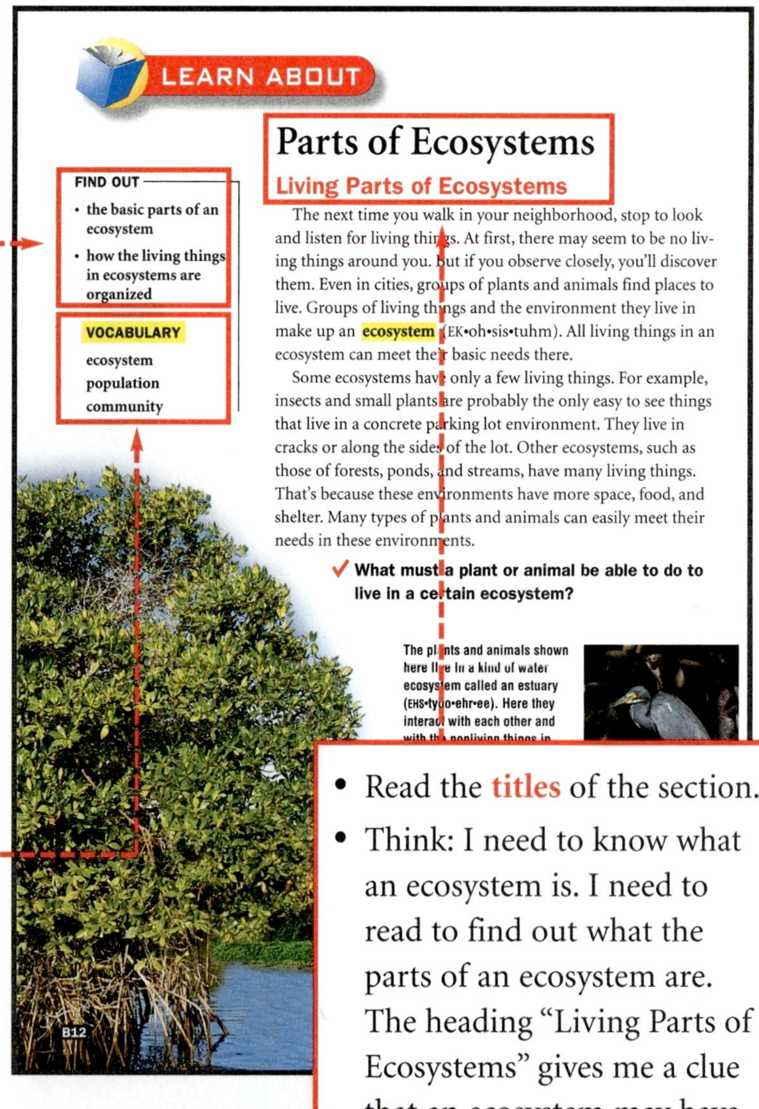

LEARN ABOUT

FIND OUT
- the basic parts of an ecosystem
- how the living things in ecosystems are organized

VOCABULARY
ecosystem
population
community

Parts of Ecosystems

Living Parts of Ecosystems

The next time you walk in your neighborhood, stop to look and listen for living things. At first, there may seem to be no living things around you. But if you observe closely, you'll discover them. Even in cities, groups of plants and animals find places to live. Groups of living things and the environment they live in make up an **ecosystem** (EK•oh•sis•tuhm). All living things in an ecosystem can meet their basic needs there.

Some ecosystems have only a few living things. For example, insects and small plants are probably the only easy to see things that live in a concrete parking lot environment. They live in cracks or along the sides of the lot. Other ecosystems, such as those of forests, ponds, and streams, have many living things. That's because these environments have more space, food, and shelter. Many types of plants and animals can easily meet their needs in these environments.

✓ What must a plant or animal be able to do to live in a certain ecosystem?

The plants and animals shown here live in a kind of water ecosystem called an estuary (EHS•tyoo•ehr•ee). Here they interact with each other and

B12

- Read the **titles** of the section.
- Think: I need to know what an ecosystem is. I need to read to find out what the parts of an ecosystem are. The heading "Living Parts of Ecosystems" gives me a clue that an ecosystem may have both living and nonliving parts.

During Reading

Find the **main idea** in the first paragraph.

- Groups of living things and their environment make up an ecosystem.

Find **details** in the next paragraph that support the main idea.

- Some ecosystems have only a few living things.
- Environments that have more space, food, and shelter have many living things.
- Plants and animals in an ecosystem can meet all their basic needs in their ecosystem.

Check your understanding of what you have read.

- Answer the question at the end of the section.
- If you're not sure of the answer, reread the section and look for the answer to the question.

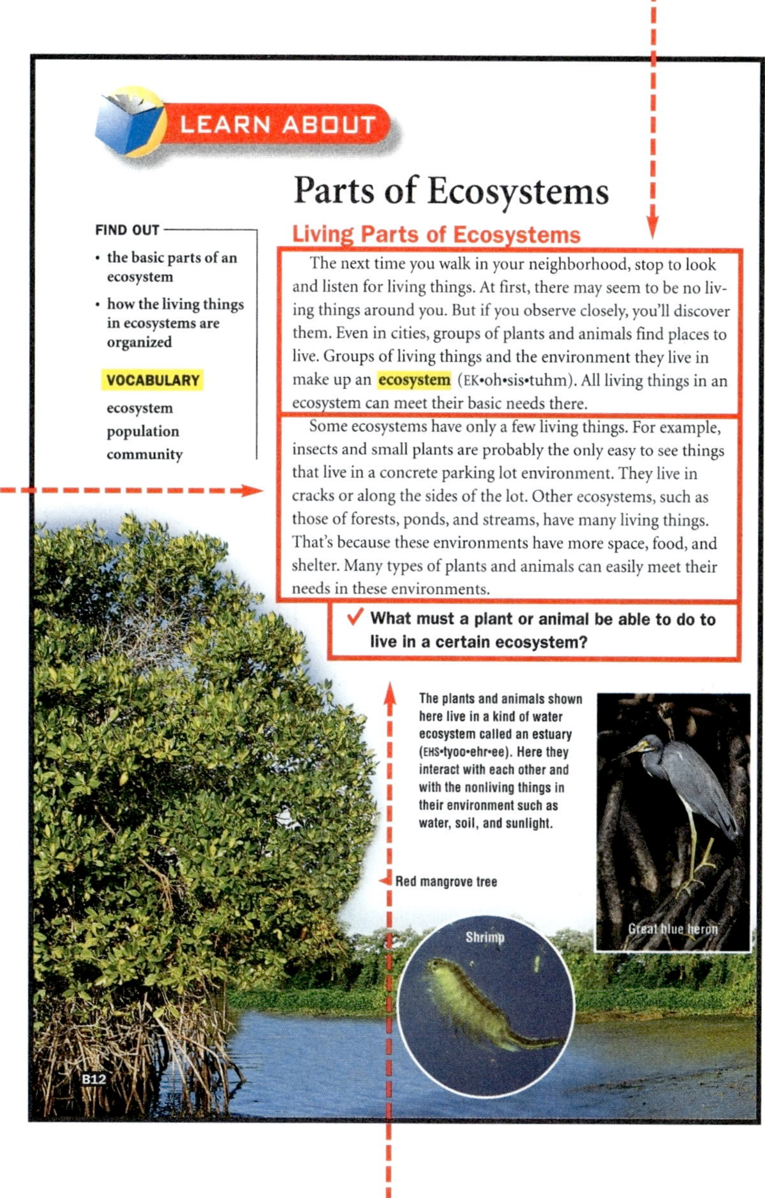

After Reading

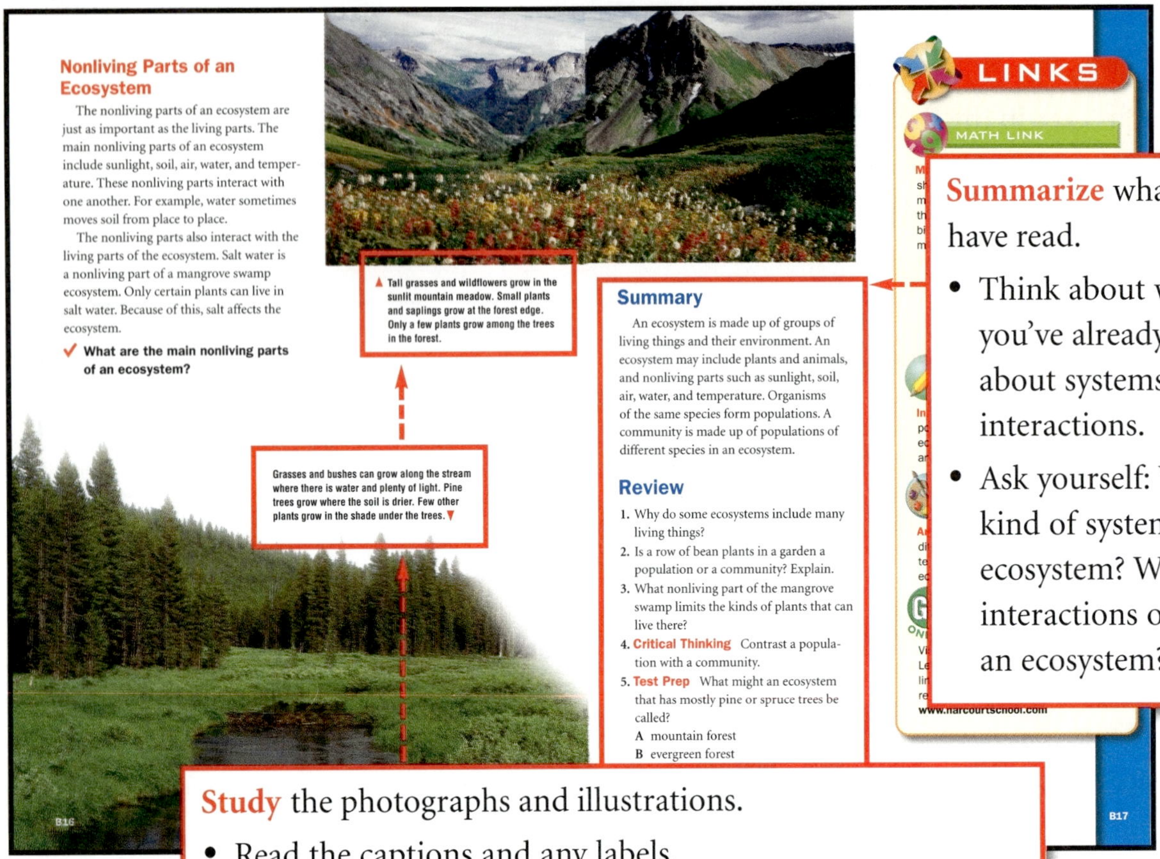

Nonliving Parts of an Ecosystem

The nonliving parts of an ecosystem are just as important as the living parts. The main nonliving parts of an ecosystem include sunlight, soil, air, water, and temperature. These nonliving parts interact with one another. For example, water sometimes moves soil from place to place.

The nonliving parts also interact with the living parts of the ecosystem. Salt water is a nonliving part of a mangrove swamp ecosystem. Only certain plants can live in salt water. Because of this, salt affects the ecosystem.

✓ **What are the main nonliving parts of an ecosystem?**

▲ Tall grasses and wildflowers grow in the sunlit mountain meadow. Small plants and saplings grow at the forest edge. Only a few plants grow among the trees in the forest.

Grasses and bushes can grow along the stream where there is water and plenty of light. Pine trees grow where the soil is drier. Few other plants grow in the shade under the trees. ▼

Summary

An ecosystem is made up of groups of living things and their environment. An ecosystem may include plants and animals, and nonliving parts such as sunlight, soil, air, water, and temperature. Organisms of the same species form populations. A community is made up of populations of different species in an ecosystem.

Review

1. Why do some ecosystems include many living things?
2. Is a row of bean plants in a garden a population or a community? Explain.
3. What nonliving part of the mangrove swamp limits the kinds of plants that can live there?
4. **Critical Thinking** Contrast a population with a community.
5. **Test Prep** What might an ecosystem that has mostly pine or spruce trees be called?
   A mountain forest
   B evergreen forest

LINKS

MATH LINK

www.harcourtschool.com

B16 B17

Summarize what you have read.

- Think about what you've already learned about systems and interactions.
- Ask yourself: What kind of system is an ecosystem? What interactions occur in an ecosystem?

Study the photographs and illustrations.

- Read the captions and any labels.
- Think: What kind of ecosystem is shown in the photographs? What are the nonliving parts of the ecosystem? What living parts of the ecosystem are shown?

For more reading strategies and tips, see pages R38–R49.

Reading about science helps you understand the conclusions you have made based on your investigations.

Writing to Communicate

Writing about what you are learning helps you connect the new ideas to what you already know. Scientists **write** about what they learn in their research and investigations to help others understand the work they have done. As you work like a scientist, you will use the following kinds of writing to describe what you are doing and learning.

In **informative writing,** you may

- describe your observations, inferences, and conclusions.
- tell how to do an experiment.

In **narrative writing,** you may

- describe something, give examples, or tell a story.

In **expressive writing,** you may

- write letters, poems, or songs.

In **persuasive writing,** you may

- write letters about important issues in science.
- write essays expressing your opinions about science issues.

Writing about what you have learned about science helps others understand your thinking.

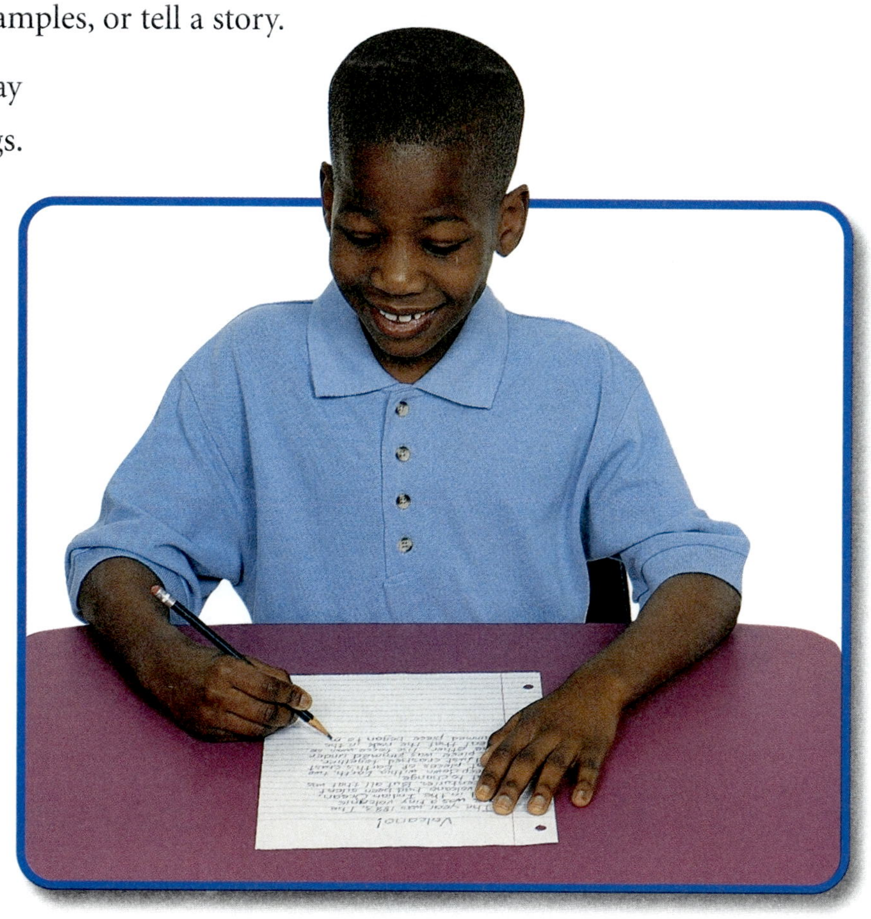

Using Numbers

Scientists **use numbers** when they collect and display their data. Understanding numbers and using them to show the results of investigations are important skills that a scientist must have. As you work like a scientist, you will use numbers in the following ways:

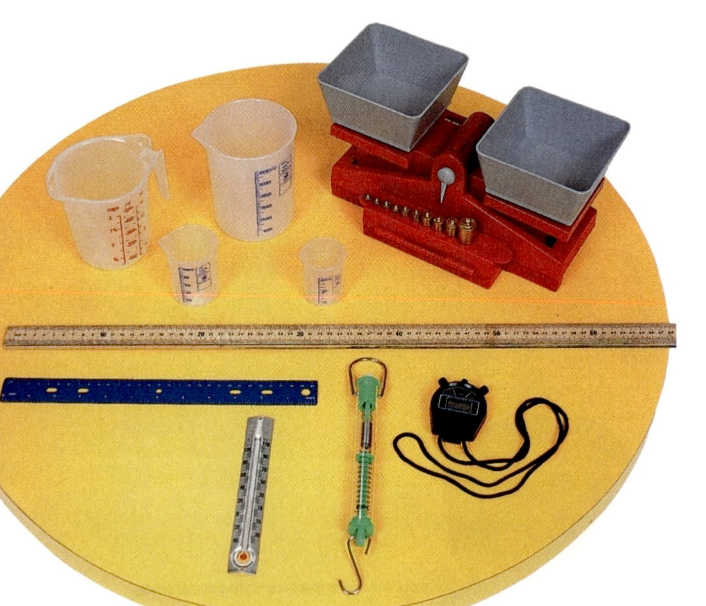

Measuring

Scientists make accurate measurements as they gather data. They use many different measuring instruments, such as thermometers, clocks and timers, rulers, a spring scale, and a balance, and they use beakers and other containers to measure liquids.

For more information about using measuring tools, see pages R2–R6.

Interpreting Data

Scientists collect, organize, display, and interpret data as they do investigations. Scientists choose a way to display data that helps others understand what they have learned. Tables, charts, and graphs are good ways to display data so that it can be interpreted by others.

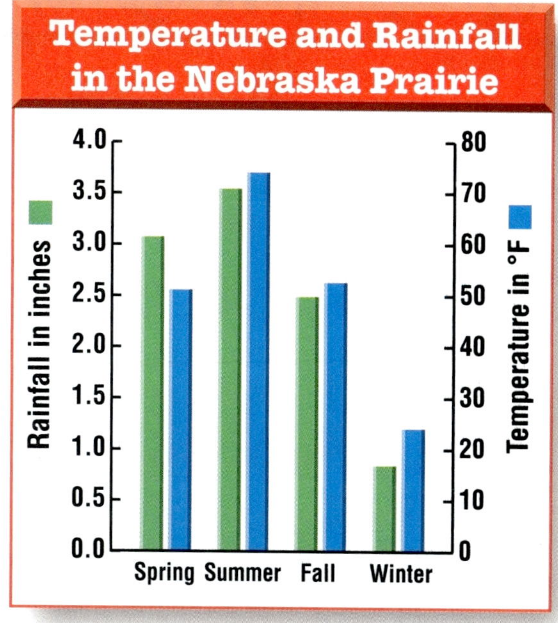

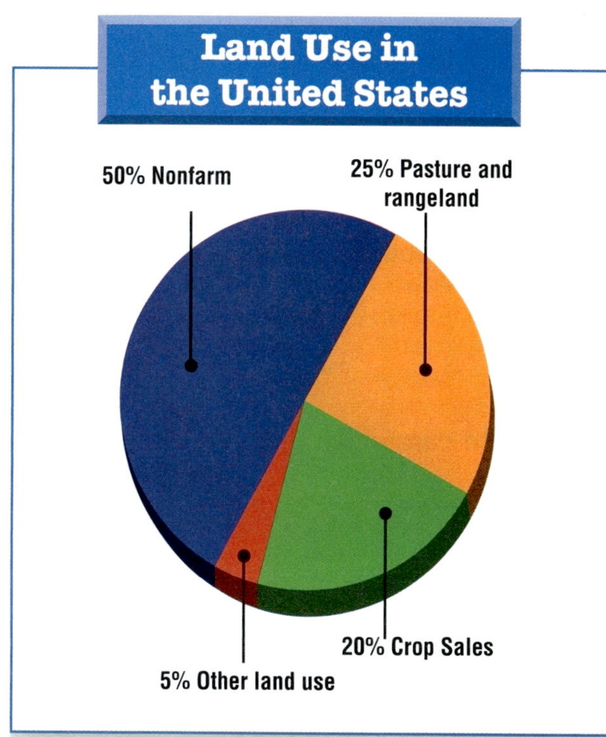

Using Number Sense

Scientists must understand what the numbers they use represent. They compare and order numbers, compute with numbers, read and understand the numbers shown on graphs, and read the scales on thermometers, measuring cups, beakers, and other tools.

Good scientists apply their math skills to help them display and interpret the data they collect.

In Harcourt Science you will have many opportunities to work like a scientist. An exciting year of discovery lies ahead!

HOW SCIENTISTS WORK

Safety in Science

Doing investigations in science can be fun, but you need to be sure you do them safely. Here are some rules to follow.

1. **Think ahead.** Study the steps of the investigation so you know what to expect. If you have any questions, ask your teacher. Be sure you understand any safety symbols that are shown.

2. **Be neat.** Keep your work area clean. If you have long hair, pull it back so it doesn't get in the way. Roll or push up long sleeves to keep them away from your experiment.

3. **Oops!** If you should spill or break something or get cut, tell your teacher right away.

4. **Watch your eyes.** Wear safety goggles anytime you are directed to do so. If you get anything in your eyes, tell your teacher right away.

5. **Yuck!** Never eat or drink anything during a science activity unless you are told to do so by your teacher.

6. **Don't get shocked.** Be especially careful if an electric appliance is used. Be sure that electric cords are in a safe place where you can't trip over them. Don't ever pull a plug out of an outlet by pulling on the cord.

7. **Keep it clean.** Always clean up when you have finished. Put everything away and wipe your work area. Wash your hands.

In some activities you will see these symbols. They are signs for what you need to act safely.

Be especially careful.

Wear safety goggles.

Be careful with sharp objects.

Don't get burned.

Protect your clothes.

Protect your hands with mitts.

Be careful with electricity.

UNIT D EARTH SCIENCE

Patterns on Earth and in Space

UNIT EXPERIMENT

Clouds and Weather Prediction

Earth's atmosphere affects you every day. You want to know how hot or cold it is and if it will rain or snow. While you study this unit, you can conduct a long-term experiment about predicting weather. Here are some questions to think about. How can you use clouds to predict the weather? Is there a pattern of changes that goes with certain clouds? What weather conditions are associated with each cloud type? Plan and conduct an experiment to find answers to these or other questions you have about patterns on Earth and in space. See pages x–xix for help in designing your experiment.

CHAPTER

1 Weather Conditions

Vocabulary Preview

atmosphere
air pressure
troposphere
stratosphere
greenhouse effect
air mass
front
cirrus
cumulus
cumulonimbus
stratus
barometer
humidity
hygrometer

Everyone talks about the weather, but weather forecasters get paid to talk about it. Many people depend on weather forecasts to plan their day. Sometimes forecasts of severe weather can even save lives.

Fast Fact

Right now, 2000 thunderstorms are happening around the Earth. While you are reading this sentence, lightning will strike the Earth about 500 times!

Fast Fact

Never underestimate the power of a flood! Just fifteen centimeters (about 6 in.) of fast-moving water can knock you off your feet. Sixty centimeters (about 2 ft) of water can carry away an automobile!

Fast Fact

If you're a famous baseball player, your number will be retired. If you're a really destructive hurricane, your name will be retired for at least 10 years! Instead of a batting average or ERA, a hurricane has a scale rating. Here's how it works:

Saffir-Simpson Hurricane Scale

Category	Wind Speed (kilometers per hour)	Wind Speed (miles per hour)
1	119–153	74–95
2	154–177	96–110
3	178–209	111–130
4	210–249	131–155
5	over 249	over 155

LESSON 1

What Makes Up Earth's Atmosphere?

In this lesson, you can . . .

INVESTIGATE a property of air.

LEARN ABOUT Earth's atmosphere.

LINK to math, writing, art, and technology.

A Property of Air

Activity Purpose Everything around you is matter. Matter is anything that takes up space and has weight. In this investigation you will **observe** a property of air. Then you will **infer** whether air is matter.

Materials

- metric ruler
- piece of string about 80 cm long
- scissors
- 2 round balloons (same size)
- safety goggles
- straight pin

Activity Procedure

1. Work with a partner. Use the scissors to carefully cut the string into three equal pieces. CAUTION **Be careful when using scissors.**
2. Tie one piece of the string to the middle of the ruler.

◄ Oxygen is part of the air you breathe. High on a mountain the particles of air are far apart. The climber can't get enough oxygen from the air. He needs extra oxygen from a tank to keep his body working properly.

3. Blow up the balloons so they are about the same size. Seal the balloons. Then tie a piece of string around the neck of each balloon.

4. Tie a balloon to each end of the ruler. Hold the middle string up so that the ruler hangs from it. Move the strings so that the ruler is balanced. (Picture A)

5. **CAUTION** **Put on your safety goggles.** Use the straight pin to pop one of the balloons. **Observe** what happens to the ruler.

Picture A

Draw Conclusions

1. Explain how this investigation shows that air takes up space.
2. Describe what happened when one balloon was popped. What property of air caused what you **observed?**
3. **Scientists at Work** Scientists often **infer** conclusions when the answer to a question is not clear or can't be **observed** directly. Your breath is invisible, but you observed how it made the balloons and the ruler behave. Even though you can't see air, what can you infer about whether or not air is matter? Explain.

Investigate Further The air around you presses on you and everything else on Earth. This property of air, called air pressure, is a result of air's weight. When more air is packed into a small space, air pressure increases. You can feel air pressure for yourself. Hold your hands around a partly filled balloon while your partner blows it up. Describe what happens. Then **infer** which property of air helps keep the tires of a car inflated.

Process Skill Tip

Observations and inferences are different things. An **observation** is made with your senses. An **inference** is an opinion based on what you have observed and what you know about a situation.

Earth's Atmosphere

FIND OUT

- some properties of air
- the layers of the atmosphere

VOCABULARY

atmosphere
air pressure
troposphere
stratosphere

The Air You Breathe

You can live for a few days without water and for many days without food. But you can live only a few minutes without air. Nearly all living things need air to carry out their life processes. The layer of air that surrounds our planet is called the **atmosphere** (AT•muhs•feer). When compared to the size of Earth, the atmosphere looks like a very thin blanket surrounding the entire planet.

The atmosphere wasn't always as it is today. It formed millions of years ago as gases from erupting volcanoes collected around the planet. This mixture of gases would have poisoned you if you had breathed it. But bacteria and other living things used gases in this early atmosphere. They released new gases as they carried out their life processes. Over time, the gas mixture changed slowly to become the atmosphere Earth has now.

The atmosphere now is made up of billions and billions of gas particles. Almost four-fifths of these gas particles are nitrogen. Oxygen, a gas that your body uses in its life processes, makes up about one-fifth of the atmosphere. Other gases, including carbon dioxide and water vapor, make up the rest of the atmosphere.

Although you can't see all of it, a thin blanket of air called the atmosphere surrounds Earth. ▼

Plants use carbon dioxide during the process of photosynthesis. Plants give off oxygen as photosynthesis occurs. Carbon dioxide also absorbs heat energy from the sun and from Earth's surface. This helps keep the planet warm.

Like carbon dioxide, water vapor can absorb heat energy. The amount of water vapor in the air varies from place to place. Air over bodies of water usually contains more water vapor than air over land. High in the air, water vapor condenses to form clouds.

Air has certain properties. As you saw in the investigation, air takes up space and has weight. All the particles of air pressing down on the surface cause **air pressure** (PRESH•er). Air pressure changes as you go higher in the atmosphere. The picture shows what a column of air might look like. At the surface of Earth, air particles are close together. The higher you go in the atmosphere, the farther apart the air particles are. So the air pressure is less as you go higher in the atmosphere.

✔ **What is the atmosphere?**

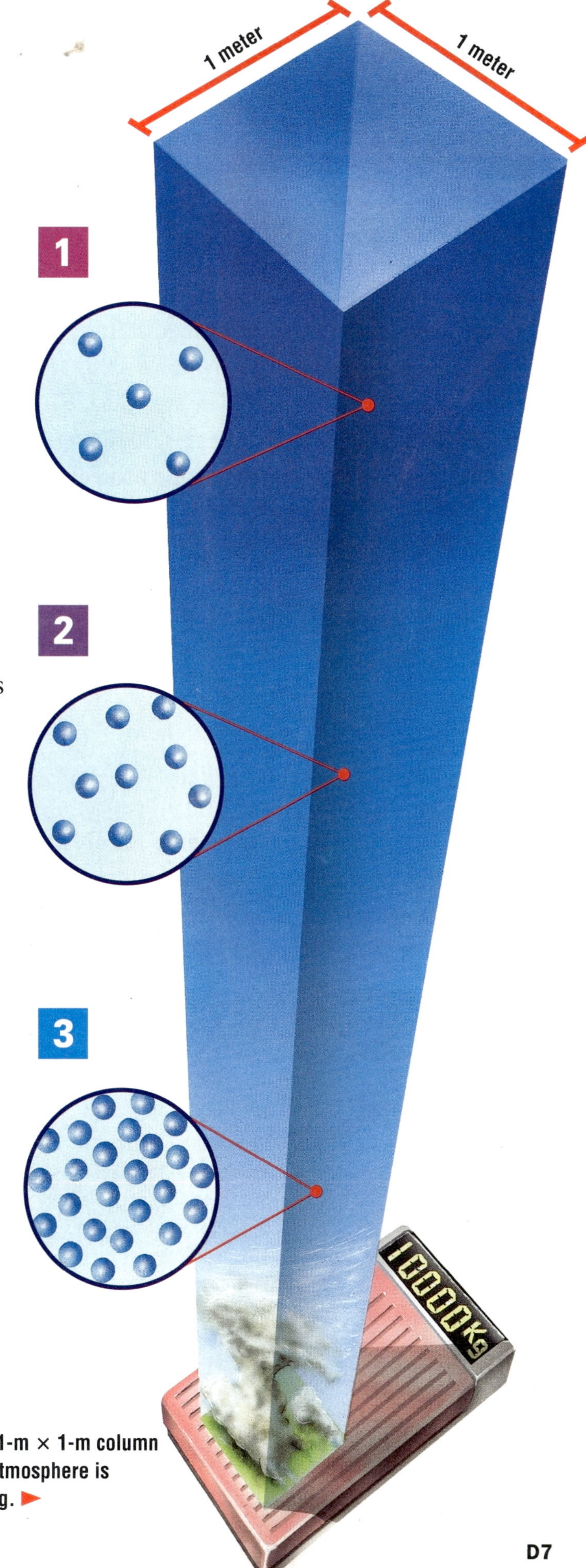

1 Air particles in the upper atmosphere have the least weight pressing on them. The particles are far apart. Air in this part of the atmosphere is much less dense than air lower in Earth's atmosphere.

2 Air near the middle of the atmosphere has more weight pressing down on it. So it is denser than air higher above Earth.

3 The weight of the entire column of air presses down on the air particles closest to Earth, forcing them close together. This makes air densest at Earth's surface. Air pressure is greatest where air is densest.

The mass of a 1-m × 1-m column of the Earth's atmosphere is about 10,000 kg. ►

Atmosphere Layers

Earth's atmosphere is divided into four layers. The layer closest to Earth is the **troposphere** (TROH•poh•sfeer). We live in the troposphere and breathe its air. Almost all weather happens in this layer. In the troposphere, air temperature decreases as you go higher.

Some airplanes that travel long distances fly in the **stratosphere** (STRAT•uh•sfeer) to be above most bad weather. The stratosphere contains most of the atmosphere's ozone, a kind of oxygen. The ozone protects living things from the sun's harmful rays. Temperatures in the stratosphere increase with height.

In the mesosphere (MES•oh•sfeer), air temperature decreases with height. In fact, the mesosphere is the coldest layer of the atmosphere. The thermosphere (THER•moh•sfeer) is the hot, outermost layer of air. In the thermosphere, temperature increases quickly with height. Temperatures high in the thermosphere can reach thousands of degrees Celsius.

✓ **What are the four layers of the atmosphere?**

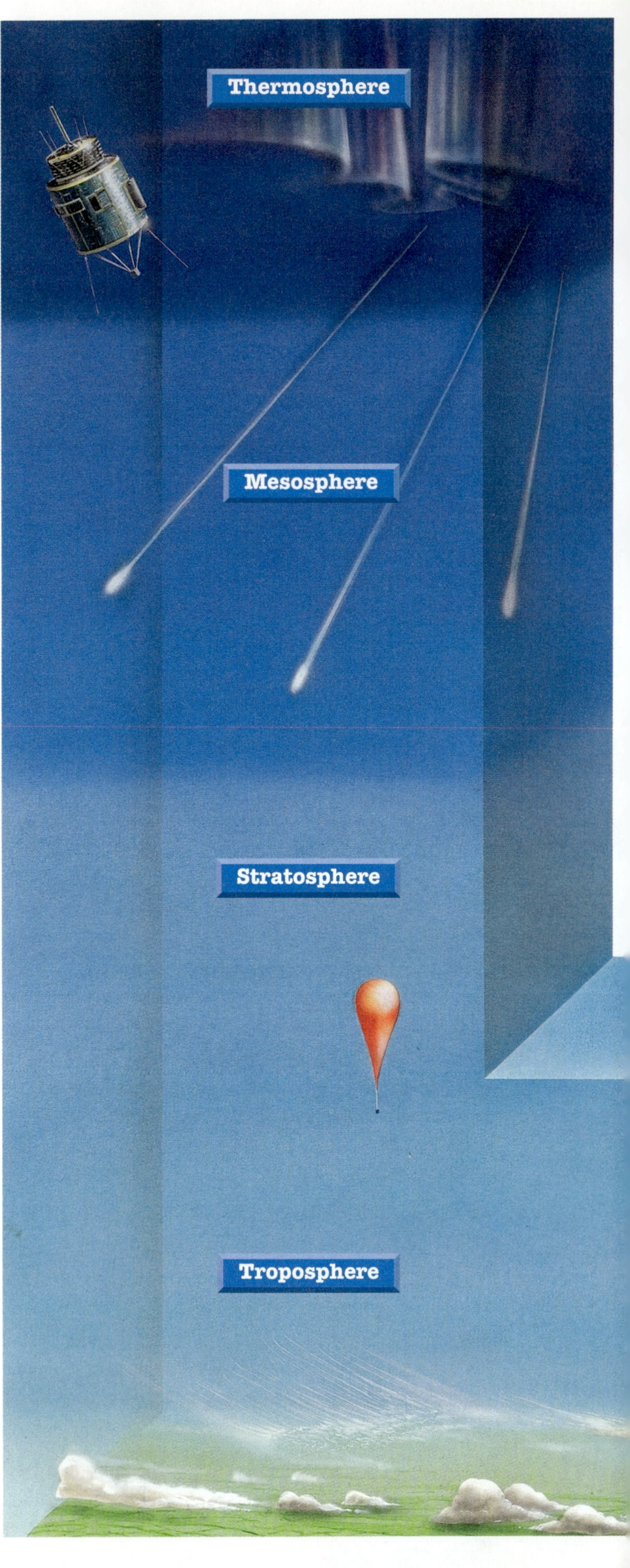

Earth's atmosphere is divided into four layers based on changes in air temperature. Each layer blends into the next. The thermosphere fades into outer space, where there is no air at all. ►

Summary

The thin blanket of air that surrounds Earth is called the atmosphere. Earth's atmosphere is divided into four layers based on changes in temperature. The layers, starting with the one closest to Earth, are the troposphere, stratosphere, mesosphere, and thermosphere.

Review

1. What is the atmosphere?
2. How does air pressure change with height?
3. How is the atmosphere divided?
4. **Critical Thinking** Compare and contrast the stratosphere and the mesosphere.
5. **Test Prep** In which layer of the atmosphere does most weather occur?
   A troposphere
   B stratosphere
   C mesosphere
   D thermosphere

MATH LINK

Solve a Two-Step Problem In the troposphere the air temperature drops about $6\frac{1}{2}$°C for every 1 kilometer increase in height. If the troposphere is about 10 kilometers thick and the air temperature at the ground is 30°C, what is the temperature at a height of 2 kilometers?

WRITING LINK

Informative Writing—Description Suppose that you are falling from space toward Earth. For your teacher, write a story describing what you see and feel as you go through each layer of the atmosphere.

ART LINK

Atmosphere Layers Paint a picture showing the atmosphere as you would see it from space. Label the layers.

TECHNOLOGY LINK

Learn more about Earth's atmosphere and weather by visiting this Internet site.

www.scilinks.org/harcourt

LESSON 2

How Do Air Masses Affect Weather?

In this lesson, you can . . .

INVESTIGATE wind speed.

LEARN ABOUT what causes weather.

LINK to math, writing, health, and technology.

Wind, which is air in motion, keeps these kites fluttering in the sky.

Wind Speed

Activity Purpose Have you ever flown a kite? A strong wind makes the kite flutter and soar through the air. A gentle breeze is usually not enough to keep the kite flying. What is wind? Wind is air in motion. In this investigation you will make an instrument to **measure** wind speed.

Materials

- sheet of construction paper
- tape
- hole punch
- 4 gummed reinforcements
- glue
- piece of yarn about 20 cm long
- strips of tissue paper, about 1 cm wide and 20 cm long

Activity Procedure

1. Form a cylinder with the sheet of construction paper. Tape the edge of the paper to keep the cylinder from opening.
2. Use the hole punch to make two holes at one end of the cylinder. Punch them on opposite sides of the cylinder and about 3 cm from the end. Put two gummed reinforcements on each hole, one on the inside and one on the outside. (Picture A)
3. Thread the yarn through the holes, and tie it tightly to form a handle loop.

Wind Scale			
Speed (km/h)	**Description**	**Objects Affected**	**Windsock Position**
0	no breeze	no movement of wind	
6–19	light breeze	leaves rustle, wind vanes move, wind felt on face	
20–38	moderate breeze	dust and paper blow, small branches sway	
39–49	strong breeze	umbrellas hard to open, large branches sway	

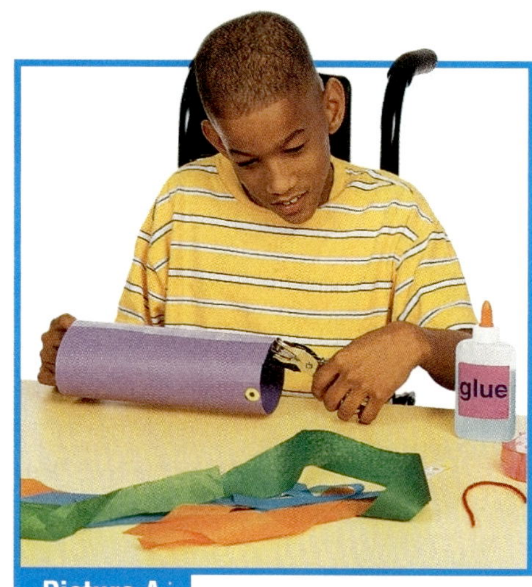

Picture A

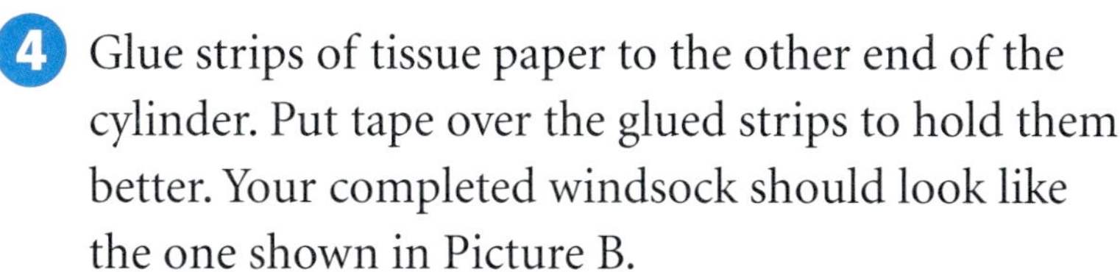

4. Glue strips of tissue paper to the other end of the cylinder. Put tape over the glued strips to hold them better. Your completed windsock should look like the one shown in Picture B.

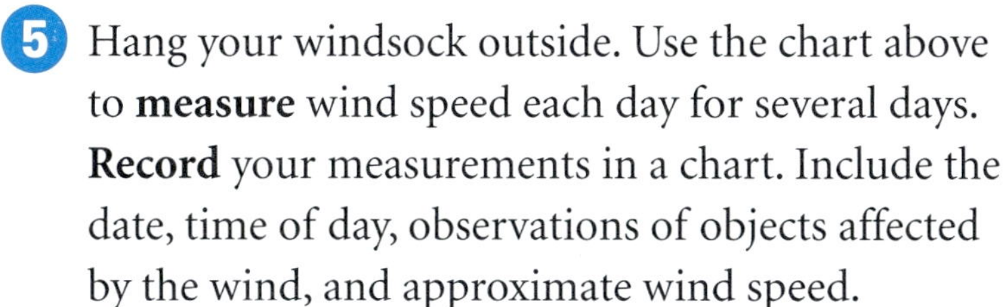

5. Hang your windsock outside. Use the chart above to **measure** wind speed each day for several days. **Record** your measurements in a chart. Include the date, time of day, observations of objects affected by the wind, and approximate wind speed.

Picture B

Draw Conclusions

1. How fast was the weakest wind you **measured**? How fast was the strongest wind?
2. How did you determine the speed of the wind?
3. **Scientists at Work** *Light, moderate,* and *strong* are adjectives describing wind speed. Scientists often **use numbers** to describe things because, in science, numbers are more exact than words. What is the wind speed measurement in kilometers per hour if the wind is making large tree branches sway?

Investigate Further Use a magnetic compass to determine which way is north from your windsock. **Measure** both wind speed and direction each day for a week. **Record** your data in a chart.

Process Skill Tip

The **use of numbers** allows people to communicate precisely. Telling someone there is a gentle breeze is not as exact as saying the wind is blowing at 8 km/h.

Air and Weather

FIND OUT

- how the sun affects weather
- what makes an air mass

VOCABULARY

greenhouse effect
air mass
front
cirrus
cumulus
cumulonimbus
stratus

Air and the Sun

Have you ever watched a weather report on television? If so, you know that temperature, air pressure, and wind are some of the things reported. You also know that these weather conditions change every day. But do you know why?

Weather begins with the sun, which provides energy for making weather. But the amount of the sun's energy reaching Earth is not the same everywhere. More energy reaches the equator than the poles. This uneven heating is part of what causes air to move and what makes weather.

Most of the sun's energy never reaches Earth. It is lost in space. Of the tiny fraction of the sun's energy that does reach Earth, about three-tenths is reflected out into space. Another three-tenths warms the air. The other four-tenths warms the land and oceans. The atmosphere traps this heat much like the glass of a greenhouse. Without this **greenhouse effect**, Earth would reflect most of the sun's energy back into space and Earth's surface would be too cold to support life.

✓ **How does the atmosphere work like a greenhouse?**

▲ Have you ever seen ripples like these above a paved road on a hot day? The air just above the hot pavement is also hot. Light travels differently through hot air. That's why the view is blurry.

Sunlight passes through the atmosphere and warms Earth's surface. The greenhouse effect keeps most of the heat from escaping back into space. ▼

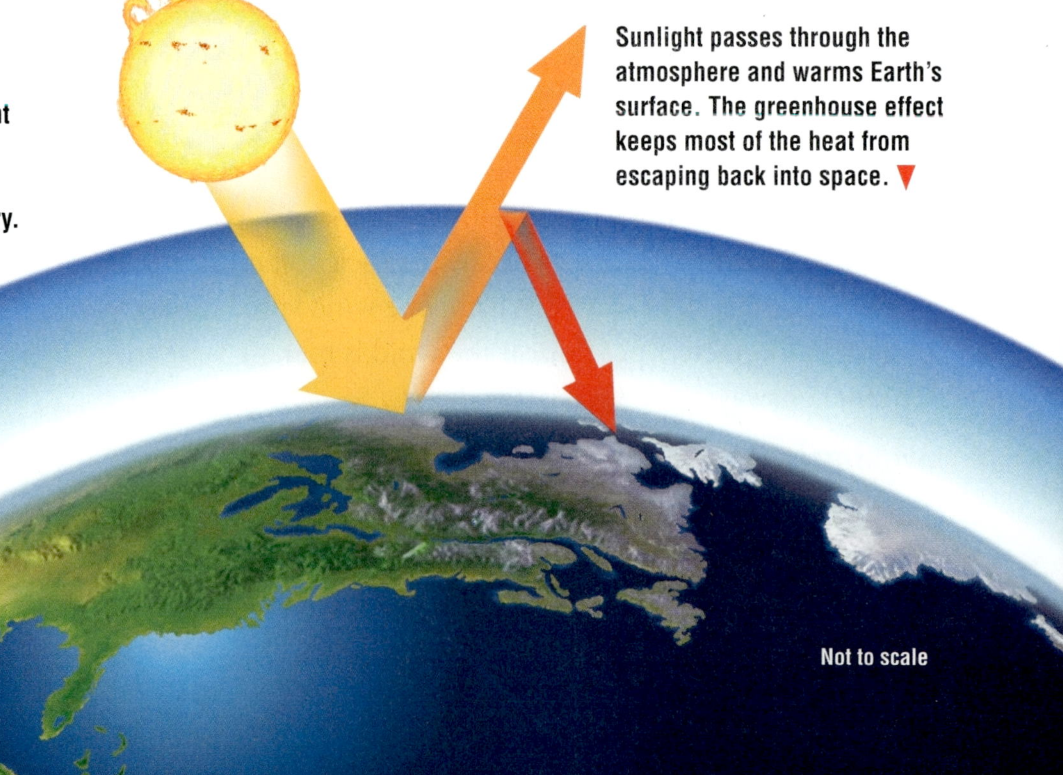

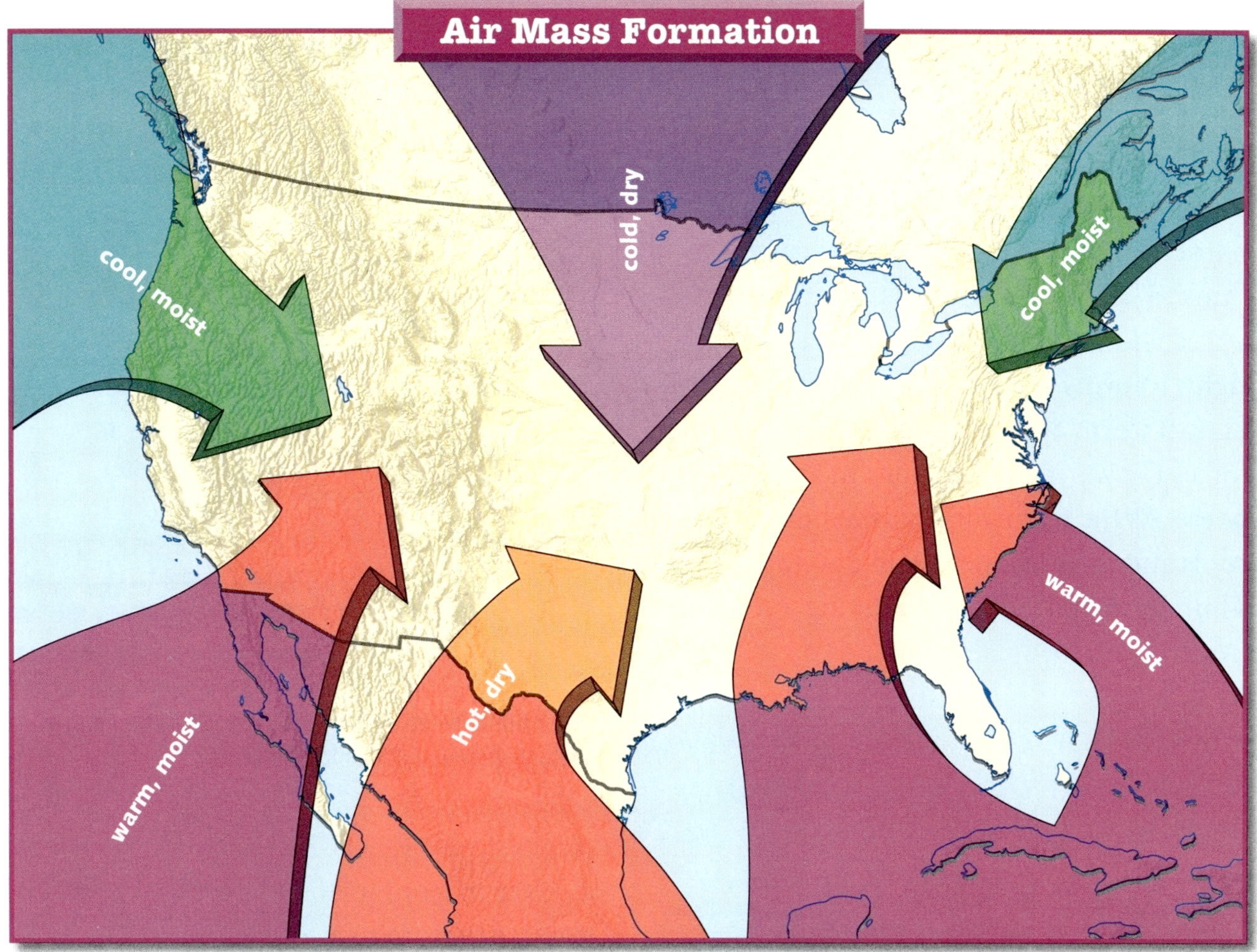

▲ Air masses form over both land and water. The map shows where the air masses that affect North America form. Cool air masses are in blue colors. Warm air masses are in red colors.

Air Masses

If you could see the air around Earth from outer space, you would see large clumps of it forming, moving over Earth's surface, and slowly changing. These huge bodies of air, which can cover thousands of kilometers, are called air masses.

Like air heated by a hot road, an **air mass** has the same general properties as the land or water over which it forms. Two properties—moisture content and temperature—are used to describe air masses. Moist air masses form over water. Air masses that form over land are generally dry. Air masses that form near Earth's poles are cold. Air masses that form in the tropics, or areas near the equator, are warm.

The map shows air masses forming and moving over the North American continent. You can see a polar air mass bringing cold, dry air from the north into the United States. You can also see warm, moist air coming in from the south as part of tropical air masses.

✓ **What is an air mass?**

Air Masses Meet

Look again at the map on page D13. What do you think happens when different air masses meet? When two air masses meet, they usually don't mix. Instead they form a border called a **front**. Most of what you think of as weather happens along fronts.

A cold front is shown on the map to the right. It forms when a cold air mass catches up to a warm air mass. The colder air mass forces the warmer air up into the atmosphere. As the warm air is pushed upward, it cools and forms clouds. Rain develops. Thunderstorms often occur along a cold front.

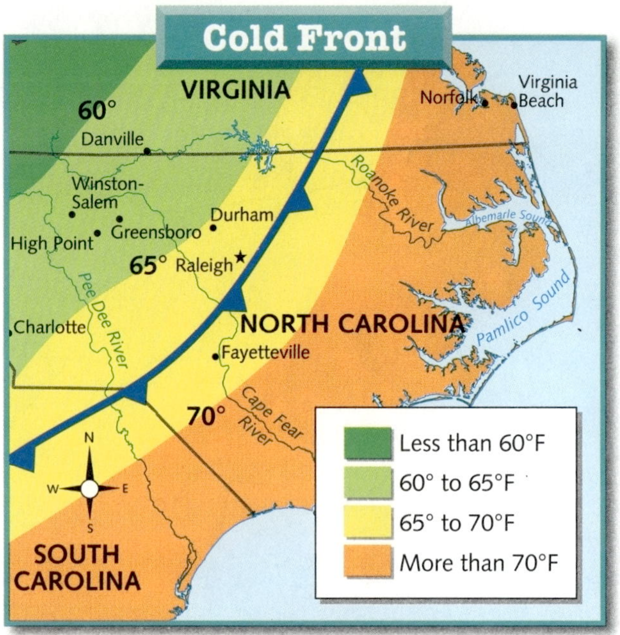

▲ **A line with triangles is the symbol for a cold front. The air is colder behind a cold front than ahead of it. The triangles point in the direction of movement. In which direction is this front moving?**

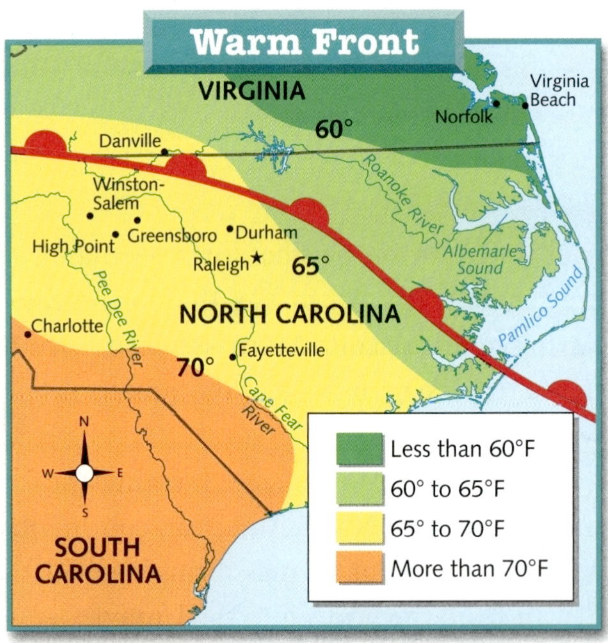

▲ **A line with half-circles is the symbol for a warm front. The half-circles point in the direction the front is moving. The air is warmer behind this front than ahead of it.**

A warm front forms when a warm air mass catches up to a cold air mass. The warm air slides up over the colder, denser air. Clouds form, sometimes many miles ahead of where the front is moving along the ground. Steady rain or snow may fall as the front approaches and passes. Then the sky becomes clear of clouds and the temperature becomes higher.

Sometimes a front stops moving. Such a front is called a stationary front. A stationary front can stay in one place for several days. The constant fall of snow or rain along a stationary front can leave behind many inches of snow or cause a flood.

✔ **What is a front?**

Clouds and the Weather

If you look up in the sky right now, you will probably see clouds with different shapes. Weather scientists classify a cloud based on its shape, its color, and where it forms in the atmosphere. The clouds you see are likely to be all the same type.

Because clouds form along a front, they can help predict weather. You know that big, dark storm clouds mean rain. But did you know that cirrus, or "horsetail," clouds can mean cooler, fair weather? Read the Inside Story to learn more about clouds and weather prediction.

✔ **How do weather scientists classify clouds?**

THE INSIDE STORY

Clouds

Learning about clouds and how they form can help you predict the weather. Read the captions below to learn more.

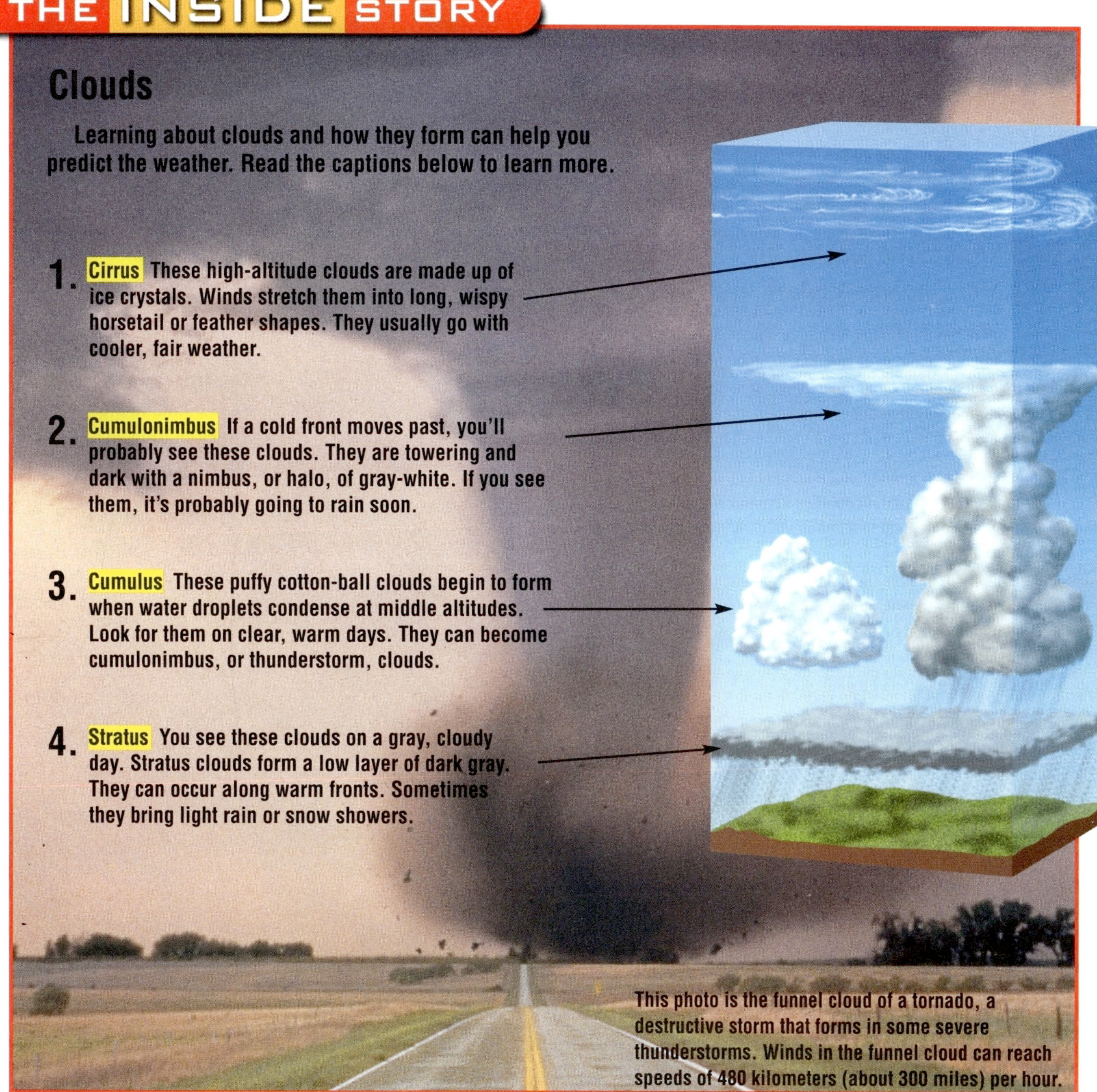

1. **Cirrus** These high-altitude clouds are made up of ice crystals. Winds stretch them into long, wispy horsetail or feather shapes. They usually go with cooler, fair weather.

2. **Cumulonimbus** If a cold front moves past, you'll probably see these clouds. They are towering and dark with a nimbus, or halo, of gray-white. If you see them, it's probably going to rain soon.

3. **Cumulus** These puffy cotton-ball clouds begin to form when water droplets condense at middle altitudes. Look for them on clear, warm days. They can become cumulonimbus, or thunderstorm, clouds.

4. **Stratus** You see these clouds on a gray, cloudy day. Stratus clouds form a low layer of dark gray. They can occur along warm fronts. Sometimes they bring light rain or snow showers.

This photo is the funnel cloud of a tornado, a destructive storm that forms in some severe thunderstorms. Winds in the funnel cloud can reach speeds of 480 kilometers (about 300 miles) per hour.

Air Masses Move

You can see air masses moving from place to place by watching how weather forms and changes. In the investigation you built a device to measure wind speed. Wind speed often increases as a front approaches. Wind direction also changes.

Air pressure also changes as air masses move over an area. As a cold front moves closer, air pressure often drops. Air pressure usually rises as the cold front moves over the area.

Temperature, too, changes as a front moves over an area. Warmer air is brought into a region by a warm front. Likewise, the temperature goes down when a cold front moves through an area.

✔ **How does air pressure change as a front moves toward and then over an area?**

The wind direction on each side of this cold front is shown by the arrows.

The arrow of a weather vane points to the direction from which the wind is blowing.

▲ The same cold front has moved and changed shape. The wind direction changes with the shape of the front.

Summary

The sun provides the energy to make weather. The atmosphere traps heat near Earth's surface much as a greenhouse does. Air masses form over continents and oceans. When two air masses meet, they form a front. Fronts are the areas where most weather happens.

Review

1. What is the greenhouse effect?
2. What is a weather front?
3. How does a cold front form?
4. **Critical Thinking** Why are weather forecasts sometimes incorrect?
5. **Test Prep** What kind of front forms when a warm air mass catches up to a cold air mass?

   A a warm front
   B a cold front
   C a rain front
   D a hot front

LINKS

MATH LINK

Subtract Decimals You're using a rain gauge to measure rainfall in your area. You recorded a rainfall of 0.3 inch but forgot to empty the gauge. The next time you checked, the gauge read 1.5 inches. How much new rain fell?

WRITING LINK

Expressive Writing—Poem Use what you've learned in this chapter so far. For a third-grade student, write a short poem about weather. Use these words in your poem: *air mass, cold front, warm front, rain,* and *clouds.*

HEALTH LINK

Severe-Weather Safety Find out what types of severe weather most often affect your area. Find out how to be prepared for severe weather. Also find out what to do to stay safe during severe weather. Write a safety checklist, and share it with your family.

TECHNOLOGY LINK

Learn more about forecasting severe weather by viewing *Tornado Tracking* on the **Harcourt Science Newsroom Video.**

LESSON 3

How Is Weather Predicted?

In this lesson, you can . . .

INVESTIGATE how to measure air pressure.

LEARN ABOUT weather prediction.

LINK to math, writing, drama, and technology.

Air Pressure

Activity Purpose You've learned that air pressure is the force with which the atmosphere presses down on Earth. You've also learned that air pressure changes as weather changes. In this investigation you will make a barometer, an instrument to **measure** air pressure.

Materials

- safety goggles
- scissors
- large, round balloon
- plastic jar
- large rubber band
- tape
- wooden craft stick
- small index card
- ruler

CAUTION

Activity Procedure

1. CAUTION **Put on your safety goggles. Be careful when using scissors.** Use the scissors to carefully cut the neck off the balloon.
2. Have your partner hold the jar while you stretch the balloon over the open end. Make sure the balloon fits snugly over the jar. Secure the balloon with the rubber band.
3. Tape the craft stick to the top of the balloon as shown. Make sure that more than half of the craft stick stretches out from the edge of the jar. (Picture A)

◀ Rain falls from clouds when water droplets in the clouds become big and heavy.

Picture A

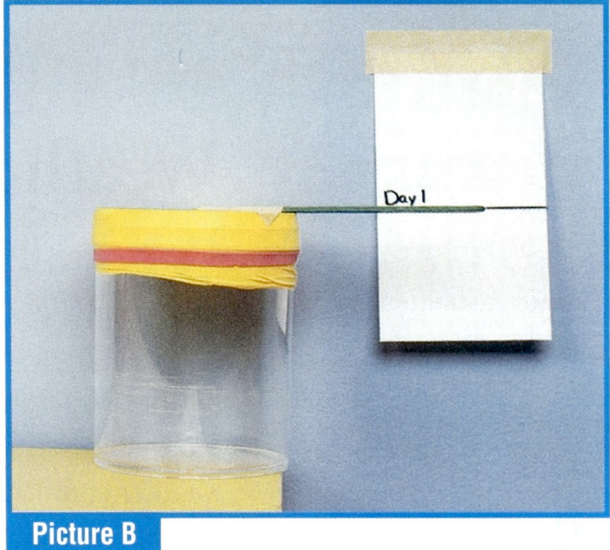

Picture B

4. On the blank side of the index card, use a pencil and a ruler to make a thin line. Label the line *Day 1.* Tape the card to a wall. Make sure the line is at the same height as the wooden stick on your barometer. (Picture B)

5. At the same time each day for a week, **measure** relative air pressure by marking the position of the wooden stick on the index card. Write the correct day next to each reading.

Draw Conclusions

1. Describe how air pressure changed during the time that you were **measuring** it.
2. What might have caused your barometer to show little or no change during the time you were taking **measurements?**
3. **Scientists at Work** Meteorologists are scientists who use instruments to **measure** weather data. How did your barometer measure air pressure?

Investigate Further Use your air pressure **measurements** and information from daily weather reports to **predict** the weather in your area for the next few days.

Process Skill Tip

When you compare data to a standard, you are **measuring**. Careful measurements can help you make inferences or draw conclusions about your data.

Weather Prediction

FIND OUT

- about instruments used to measure weather conditions
- how to read weather maps

VOCABULARY

barometer
humidity
hygrometer

Measuring Weather

Meteorologists (mee•tee•uhr•AHL•uh•jihsts) are scientists who study and measure weather conditions. Some of these conditions are air temperature, air pressure, and wind speed and direction. Meteorologists have developed tools for measuring each of these weather conditions.

When you want to know if it's hot or cold outside, you look at a thermometer or listen to a weather report. Thermometers measure the temperature of air. In the investigation, you measured another property of the atmosphere, air pressure. Air pressure is measured with an instrument called a **barometer** (buh•RAHM•uht•er). In one type of barometer, air presses down on the instrument, causing a needle to move. The needle points to a number that tells how much the air is pressing down. In other words, the instrument measures the weight of the air above it. You learned in Lesson 2 that the particles of air are closer together in a cold air mass than in a warm air mass. Most cold air masses are denser and so have higher air pressure than warm air masses.

This weather instrument is a rotating-drum barometer. It continuously records changes in air pressure. ▼

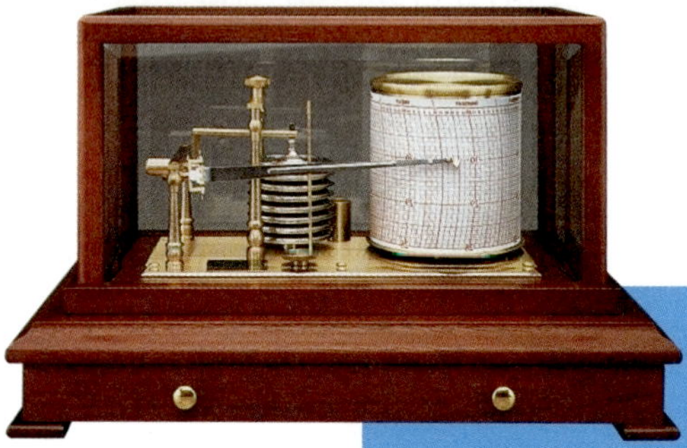

Earth's land surfaces heat faster than its bodies of water. So the air above land is usually warmer. That means it is also less dense. ▼

Evaporating water adds more humidity to the air. Air masses that form over water have more moisture than those that form over land.

This type of **hygrometer** (hy•GRAHM•uht•uhr) measures moisture in the air. The bulb of one thermometer is covered with a wet cloth. Then both thermometers are whirled in the air. The drier the air, the faster the water on the cloth evaporates (ee•VAP•uh•raytz), or dries up. This evaporation cools the cloth-covered thermometer. The temperatures of the wet and dry thermometers are compared to find the humidity.

Another characteristic of weather you can measure is **humidity** (hyoo•MID•uh•tee), or the amount of water vapor in the air. Humidity depends on several things. The area over which an air mass forms affects its humidity. For example, air masses that form over bodies of water have more moisture than air masses that form over land.

Temperature also affects how much moisture can be in the air. Warm air can have more water vapor than cool air. This is why water drops form on the outside of a glass of cold water during a warm day. Air near the glass cools. The air can no longer have as much water. The water vapor comes out of the air, forming drops.

In the investigation in Lesson 2, you made a windsock. With it you estimated the speed and direction of the wind. Meteorologists measure wind speed by using an instrument called an anemometer (an•uh•MAHM•uht•er). They find wind direction by using a weather vane or windsock.

✓ What affects the humidity of air?

This is a type of anemometer that also includes a weather vane. Wind speed is measured by counting how many complete turns the cups make in one minute. Usually, a machine counts the turns.

Daily Temperature Data for Weather City, Any State			
	Daily High/ Daily Low (°F)	Record High/ Record Low (°F)	Daily Average Temperature (°F)
Sunday	89/72	101/44	81
Monday	90/74	100/50	82
Tuesday	85/65	99/42	75
Wednesday	88/69	103/40	79
Thursday	92/70	98/41	81
Friday	75/60	99/45	68
Saturday	79/62	102/44	71

▲ Charts like this are used to record daily weather conditions.

▲ Symbols on a weather map stand for fronts and weather conditions.

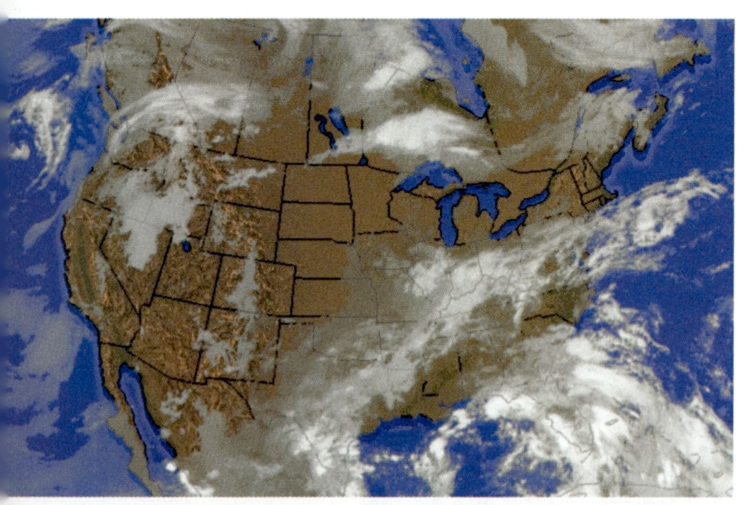

▲ A satellite photograph shows the positions of clouds and fronts. This satellite photograph matches the weather map above.

Mapping and Charting Weather

By watching and measuring weather conditions, scientists can keep track of moving air masses. Scientists record their measurements on charts and maps. Then they analyze the data to predict weather.

A weather map can be big or small. Large maps show how the weather differs across a country. Small maps show weather changes across a state or a smaller area. Satellite pictures and maps often show clouds and weather for a large part of Earth.

A weather map uses symbols to show weather conditions. Long lines marked with half-circles or triangles stand for fronts. Words or symbols describe the weather in an area. For example, small dashes may stand for rain, and small stars may stand for snow. Symbols also may show the type of clouds that are in the area.

✔ **How do weather charts and maps help scientists predict the weather?**

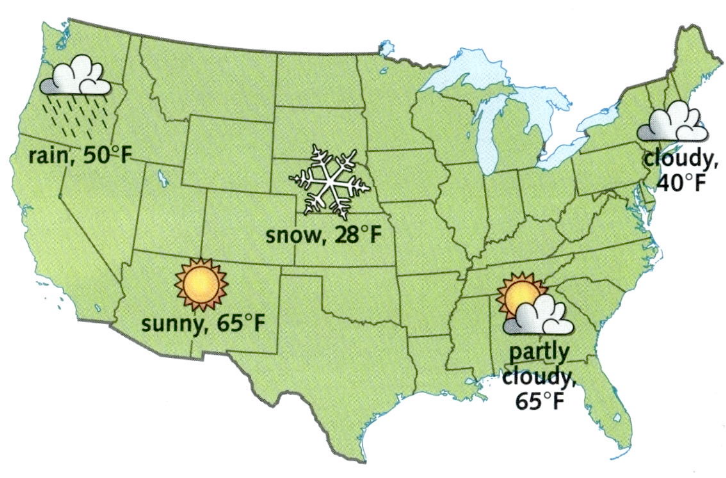

▲ Simple weather maps like this are often printed in daily newspapers. What are the temperature and the weather in Nebraska?

Summary

Meteorologists are scientists who study and measure weather conditions. These conditions include air pressure, air temperature, humidity, and wind speed and direction. By measuring and studying weather conditions, meteorologists are able to predict the weather.

Review

1. What is a hygrometer?
2. What factors affect humidity?
3. What are two tools that meteorologists use to study and predict the weather?
4. **Critical Thinking** An air mass forms over Alaska. Describe what you think the temperature and humidity of this air mass will be like.
5. **Test Prep** Which instrument is used to measure air pressure?
   A thermometer
   B weather vane
   C hygrometer
   D barometer

LINKS

MATH LINK

Subtract Whole Numbers Make a chart like the one on page D22. In it, list the daily high and low temperatures of your area for one week. Find how much the temperature changed each day.

WRITING LINK

Expressive Writing—Friendly Letter Suppose you have a pen pal who lives in an area of the country very different from your area. Write a letter to your pen pal. Describe how the weather in your area changes when a cold front moves through.

DRAMA LINK

Be a Weather Forecaster Make a weather map showing imaginary weather conditions for your state. Present your forecast to the rest of the class. Make your presentation more interesting by using props.

TECHNOLOGY LINK

Learn more about tools for measuring weather conditions by joining a *Tornado Chase* on **Harcourt Science Explorations CD-ROM.**

Red Sprites, Blue Jets, and E.L.V.E.S.

Suppose you saw something completely new. How would you describe it? That was the challenge facing some airplane pilots and scientists. They tried to name unusual flashes they saw in the sky by calling them "upward lightning," "flames," and even "giant glowing doughnuts"!

Red sprite

T 04:00:20.00 UAF

Sprites, Jets, and ELVES

In 1989 scientists began trying to show that these light flashes are real. A videotape showed the unusual flashes during a thunderstorm. Nearly 20 were photographed during the early 1990s using video cameras that could work with very little light.

At least three types of flashes were identified. The scientists finally decided to name them sprites, jets, and ELVES.

ELVES stands for "*e*missions of *l*ight and *v*ery-low-frequency perturbations from *e*lectromagnetic-pulse *s*ources." You can see why the term is abbreviated. ELVES are very dim, quick red flashes. They move outward like ripples on a pond.

Sprites are red and seem to move in groups. They appear above a thunderstorm system, 65–75 kilometers (about 40–47 mi) above the ground. They have a "head" and strands coming down from the head. Sprites can occur over both sea and land.

Blue jets were once described as rocket lightning. Not until 1994 did weather scientists show, by using color video, that these glowing streaks are blue. Blue jets occur lower in the atmosphere than red sprites do, at 40–50 kilometers (about 25–30 mi) above the ground. Blue jets travel around 100 kilometers (about 62 mi) per second.

People in airplanes and on mountains could see blue jets and red sprites because the people were above storms. No one had ever seen ELVES because they last only a thousandth of a second. This is far too little time for the human eye to see. In 1990, videos taken from the space shuttle did show ELVES. Five more years passed before a second video of ELVES was made.

Finding Sprites, Jets, and ELVES

There are several reasons why it took so long to discover these unusual lights:

- They occur only above thunderstorms, so clouds usually block the view from the ground.
- They are dim and can be seen only after the eyes have adjusted to the dark. That adjustment is spoiled by bright lightning.
- Sprites last only about 3 ten-thousandths of a second (0.0003 second).
- Only about 1 in 100 lightning strikes produces these lights.

It took careful observation to find out that sprites, jets, and ELVES are real.

THINK ABOUT IT

1. What do you think scientists thought of the early reports of sprites and jets, before the videotape?
2. Why do scientists use low-light video cameras to take pictures of sprites, jets, and ELVES?

CAREERS
METEOROLOGIST

What They Do

Meteorologists study the atmosphere and the changes that produce different kinds of weather. Meteorologists may work for business or government. They may also research new uses for computer programs in the study of weather.

Education and Training

Someone who wants to be a meteorologist must study science in college. Many meteorologists get further training in weather research and technology after college.

WEB LINK
For Science and Technology updates, visit the Harcourt Internet site.
www.harcourtschool.com

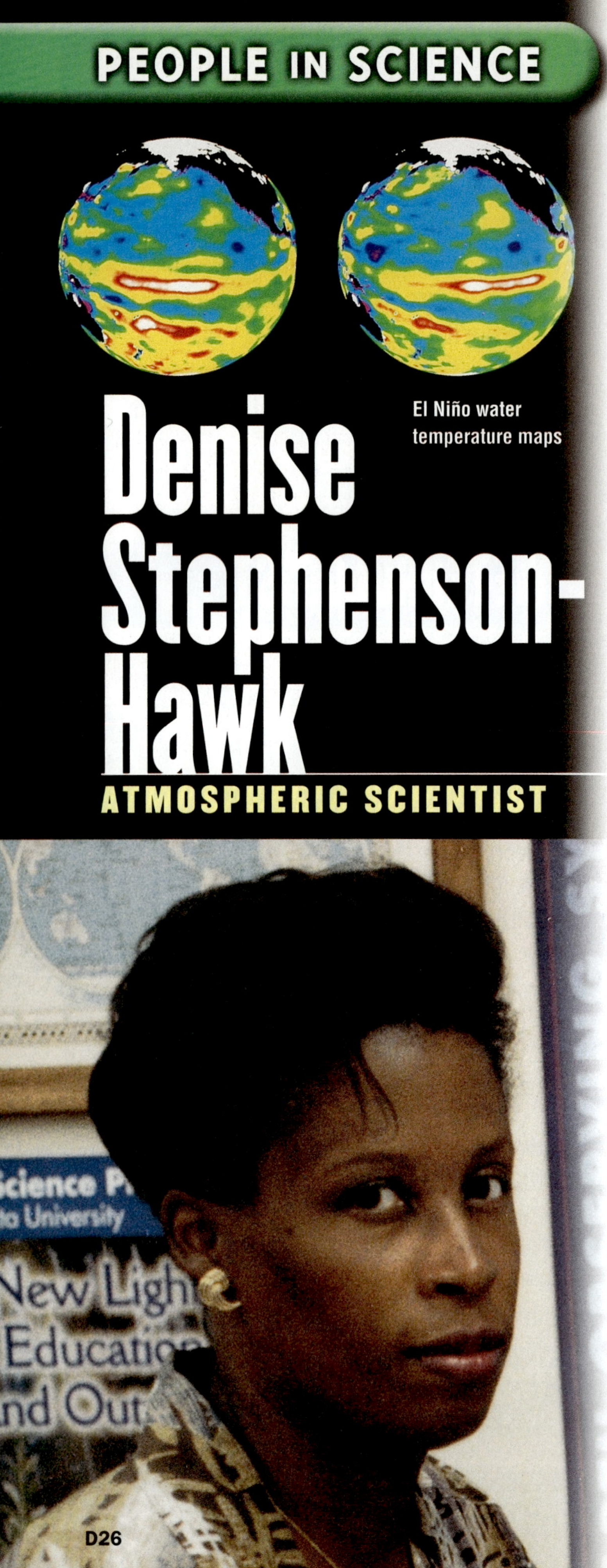

El Niño water temperature maps

Denise Stephenson-Hawk

ATMOSPHERIC SCIENTIST

Dr. Denise Stephenson-Hawk always loved school and was especially good in math. She skipped her senior year in high school and entered Spelman College. While there, she received a scholarship to a summer program at the National Aeronautics and Space Administration (NASA). She worked on a project to test panels for the space shuttle to make sure the panels would withstand the high temperatures they experience upon reentry into Earth's atmosphere. Stephenson-Hawk was so excited by what she learned about the atmosphere that she decided to apply her math skills to the study of atmospheric science.

Stephenson-Hawk's first job was at AT&T Bell Laboratories. There she made computer models to learn how sound travels in the ocean. After teaching mathematical modeling at Spelman College, Stephenson-Hawk moved to Clark Atlanta University in Georgia. There she works as a senior research scientist and associate professor of physics.

Stephenson-Hawk is also a member of the Climate Analysis Center (CAC) at the National Oceanic and Atmospheric Administration (NOAA). The CAC uses computer models to analyze and predict climate changes that happen in a short time. Stephenson-Hawk's special project has been building computer models of the effects of El Niño, a series of events set off by warmer-than-normal surface-water temperatures in the Pacific Ocean. Stephenson-Hawk and the other scientists working on this project are trying to more accurately predict the impact of El Niño so that people can better prepare for unusual weather.

THINK ABOUT IT

1. Why else might scientists be interested in studying El Niño?
2. Why do you think atmospheric scientists use computer models?

RAIN GAUGE

What is a way to measure rain and other precipitation?

Materials

- 1-liter bottle, clear plastic
- scissors
- plastic ruler
- masking tape
- wire hanger
- water

Procedure

1. **CAUTION Be careful when using scissors.** Remove the cap, and cut the top off the bottle as shown.
2. Tape the ruler to the outside of the bottle. The zero mark should be about 1 cm above the bottom.
3. Bend the coat hanger to make a basket for the bottle. Make sure the bottle will hang straight up and down.
4. Fill the bottle with water to the zero mark on the ruler. Then turn the bottle top over like a funnel and put it inside the bottle bottom.
5. Hang the rain gauge out in the open and away from any roof edges or trees— for example, on a fence.
6. After a storm, measure the rainfall and empty the bottle to the zero mark.

Draw Conclusions

How did measuring rainfall help you measure weather?

WEATHER FRONTS

How can water model a weather front?

Materials

- tall, clear jar
- hot and cold tap water
- pitcher
- food coloring
- thermometer

Procedure

1. Fill the jar halfway with cold water.
2. Fill the pitcher with hot water. Add 10 drops of food coloring.

3. Tilt the jar of cold water. Then slowly trickle the hot water down the inside of the jar. Slowly put the jar upright. Observe what happens in the jar.
4. Use the thermometer to measure the temperature of the hot water in the jar. Carefully move the thermometer down to measure the cold water in the jar. Can you find the front by using the thermometer?

Draw Conclusions

How did the hot water and cold water interact? How were they like air masses?

CHAPTER 1 Review and Test Preparation

Vocabulary Review

Use the terms below to complete the sentences. The page numbers in ( ) tell you where to look in the chapter if you need help.

atmosphere (D6)	**cirrus** (D15)
air pressure (D7)	**cumulus** (D15)
troposphere (D8)	**cumulonimbus** (D15)
stratosphere (D8)	**stratus** (D15)
greenhouse effect (D12)	**barometer** (D20)
air mass (D13)	**humidity** (D21)
front (D14)	**hygrometer** (D21)

1. The ____ is the thin layer of air that surrounds Earth.
2. The amount of water vapor in the air is called ____.
3. The warming caused when air traps some of the sun's energy is the ____.
4. ____ is the force with which the atmosphere presses down on Earth.
5. In the atmosphere, the ____ is the layer in which most weather occurs.
6. An ____ is a large body of air, and it forms and moves over land or water.
7. An instrument that measures air pressure is a ____.
8. A ____ forms when two air masses meet.
9. In the atmosphere, the layer that contains a lot of ozone is the ____.
10. An instrument that measures humidity is a ____.
11. Wispy ____ clouds are made up of ice crystals high in the atmosphere.
12. Low, gray ____ clouds form a layer and may bring light rain.
13. White, fluffy ____ clouds occur in fair weather, but they can become dark, towering ____ storm clouds.

Connect Concepts

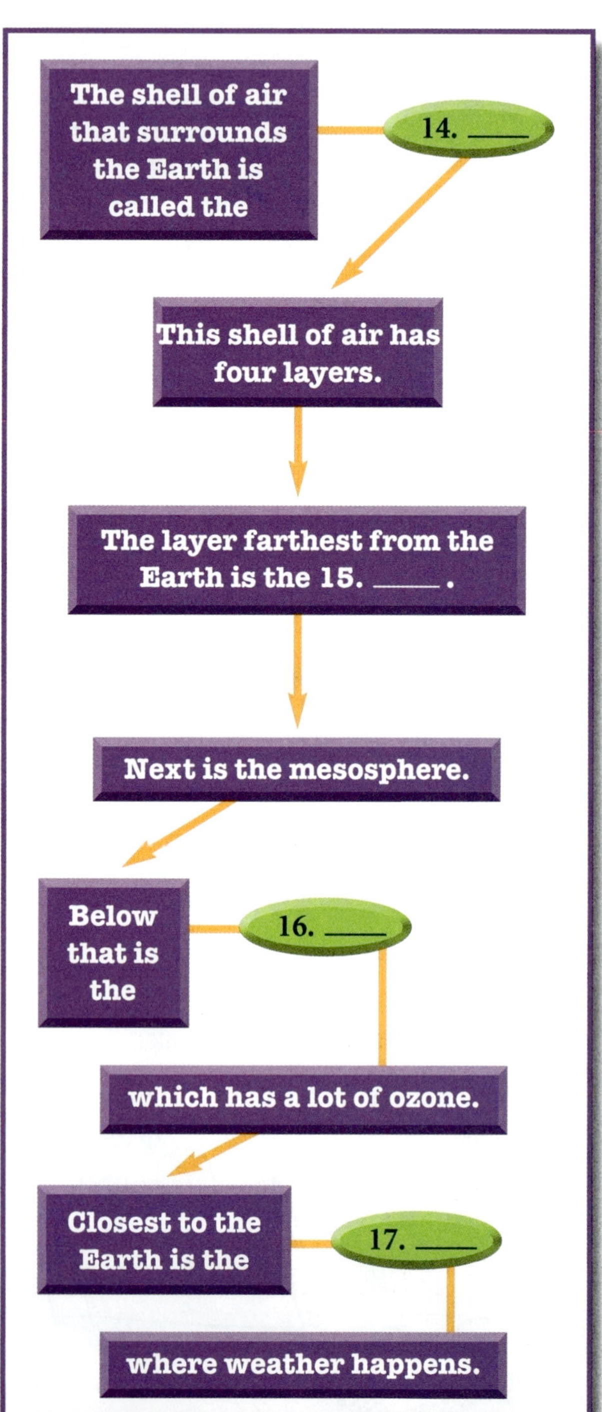

Check Understanding

Write the letter of the best choice.

18. As you get higher in the atmosphere, the space between air particles —

A decreases
B doesn't change
C increases
D masses

19. Energy from the ____ is trapped by gases in the air, causing the greenhouse effect.

F Earth
G sun
H barometer
J stratosphere

20. An air mass that forms over tropical waters would be —

A warm and moist
B cold and moist
C cold and dry
D warm and dry

21. A ____ front forms when two air masses meet and don't move.

F cold
G warm
H stationary
J pressure

22. ____ air can have more water vapor than ____ air.

A Warm, cold
B Dense, less dense
C Cold, warm
D Thermosphere, mesosphere

Critical Thinking

23. Why do mountain climbers use oxygen tanks?

24. You hear on a weather report that a cold front is coming. What weather changes can you expect?

25. Suppose you watch the weather report each day for five days. Each day the average temperature is the same and the air pressure doesn't change. What could be happening?

Process Skills Review

26. Remember the first investigation in this chapter. What did you **observe** that allowed you to **infer** that air is matter?

27. How is a **number** description of wind speed different from a word description of wind speed?

28. Suppose you will **measure** weather conditions over the next five days. What equipment will help you measure? Make a table to record your data. Include units of measure.

Performance Assessment

Weather Maps

With a partner, study the three maps your teacher gives you. Tell how the weather has changed in the map area over the past three days. Then predict what the weather will be for the next two days. Explain the reasons for your prediction.

CHAPTER 2

The Oceans

Nearly three-fourths of Earth is covered with a great body of salt water. It is always moving and full of life. Its currents bring warm temperatures to otherwise cold areas. Its depths hide great mountain ranges. And its nutrient-rich waters are home to all sorts of living things.

Vocabulary Preview

- water cycle
- evaporation
- condensation
- precipitation
- wave
- storm surge
- tide
- deep ocean current
- surface current
- shore zone
- continental shelf
- abyssal plain
- trench
- mid-ocean ridge

Fast Fact

The oceans of the Earth are vast and deep. If Earth were a smooth ball with no mountains or valleys at all, it would be completely covered with water to a depth of more than 2 kilometers (about $1\frac{1}{4}$ mi).

Fast Fact

Mount Everest

The deepest spot in the oceans is in the Mariana Trench in the Pacific—11,000 meters (about 36,000 ft) below sea level. If Mount Everest, Earth's highest mountain, were dropped into that spot, it would be covered with about $1\frac{1}{2}$ kilometers (about 1 mi) of water!

Fast Fact

The Pacific Ocean holds about half of Earth's ocean water, and it covers nearly a third of Earth's surface. Here's how three oceans compare:

Ocean Sizes

Ocean	Size (Square kilometers)	Size (Square miles)
Pacific	181,000,000	70,000,000
Atlantic	94,000,000	36,000,000
Indian	74,000,000	29,000,000

LESSON 1

What Role Do Oceans Play in the Water Cycle?

In this lesson, you can . . .

INVESTIGATE how to get fresh water from salt water.

LEARN ABOUT Earth's ocean water.

LINK to math, writing, social studies, and technology.

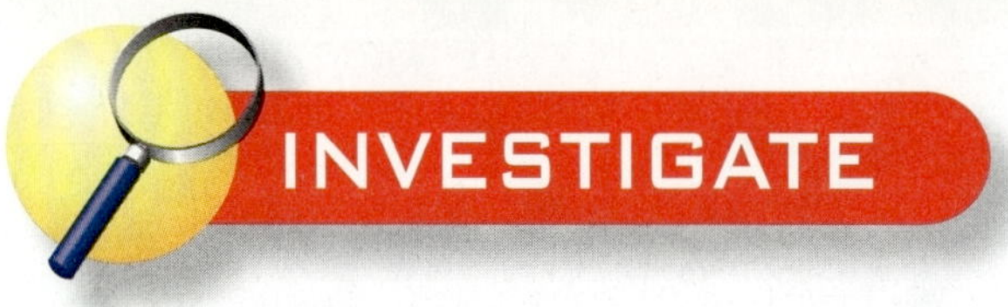

Getting Fresh Water from Salt Water

Activity Purpose If you've ever been splashed in the face by an ocean wave, you know that sea water is salty. The salt in ocean water stings your eyes, leaves a crusty white coating on your skin when it dries, and tastes like the salt you put on food. In this investigation you'll evaporate artificial ocean water to find out what is left behind. From your **observations** you will **infer** how you can get fresh water from salt water.

Materials

- container of very warm water
- salt
- spoon
- cotton swabs
- large clear bowl
- small glass jar
- plastic wrap
- large rubber band
- piece of modeling clay
- masking tape

CAUTION

Activity Procedure

1. Stir two spoonfuls of salt into the container of very warm water. Put one end of a clean cotton swab into this mixture. Taste the mixture by touching the swab to your tongue. **Record** your **observations.** CAUTION **Don't share swabs. Don't put a swab that has touched your mouth back into any substance. Never taste anything in an investigation or experiment unless you are told to do so.**

◄ Buoys float but are held in place by anchors. They mark paths where the water is deep enough for ships.

Picture A

Picture B

2. Pour the salt water into the large bowl. Put the jar in the center of the bowl of salt water. (Picture A)
3. Put the plastic wrap over the top of the bowl. The wrap should not touch the top of the jar inside the bowl. Put a large rubber band around the bowl to hold the wrap in place.
4. Form the clay into a small ball. Put the ball on top of the plastic wrap right over the jar. Make sure the plastic wrap doesn't touch the jar. (Picture B)
5. On the outside of the bowl, use tape to mark the level of the salt water. Place the bowl in a sunny spot for one day.
6. After one day, remove the plastic wrap and the clay ball. Use clean swabs to taste the water in the jar and in the bowl. **Record** your **observations.**

Draw Conclusions

1. What did you **observe** by using your sense of taste?
2. What do you **infer** happened to the salt water as it sat in the sun?
3. **Scientists at Work** The movement of water from the Earth's surface, through the atmosphere, and back to Earth's surface is called the water cycle. From what you **observed,** what can you **infer** about the ocean's role in the water cycle?

Investigate Further Put the plastic wrap and the clay back on the large bowl. Leave the bowl in the sun for several days, until all the water in the large bowl is gone. **Observe** the bowl and the jar. What can you **conclude** about ocean water?

Process Skill Tip

Observing and inferring are different things. You **observe** with your senses. You **infer**, or form an opinion, based on what you have observed and what you know about a situation.

Ocean Water

FIND OUT

- about processes that make up the water cycle
- why ocean water is salty

VOCABULARY

water cycle
evaporation
condensation
precipitation

The Water Cycle

Oceans cover more of Earth's surface than dry land does. About three-fourths of the Earth is covered by water. Almost all of that water is ocean water. Even though ocean water is salty, it provides a large amount of Earth's fresh water. Earth's water is always being recycled. As the model in the investigation showed, heat from the sun causes fresh water to evaporate (ee•VAP•uh•rayt) from the oceans, leaving the salt behind. This evaporated water condenses to form clouds. Fresh water falls from the clouds to Earth's surface as rain. This constant recycling of water is called the **water cycle**. During the cycle, water changes from a liquid to a gas and back to a liquid. The diagram on these pages shows how the water cycle works. It includes the parts played by the sun, the water, the air, and the land.

✓ **What is the water cycle?**

The sun warms the ocean, causing the water particles to move faster and faster. After a while, they have enough energy to leave the water and enter the air as water vapor. This is **evaporation**, the process by which a liquid changes to a gas.▼

A cloud forms when water vapor condenses high in the atmosphere. **Condensation** (kahn•duhn•SAY•shuhn) happens when the water vapor rises, cools, and changes from a gas to liquid water. These drops of water in a cloud are so small that they stay up in the air.

Water vapor from an ocean can be carried a long way through the atmosphere. Water that evaporates from the Gulf of Mexico may fall back to Earth's surface far away in Georgia or Virginia.

In the cloud, some water drops bump into others and stick together. The drops get bigger. When they get too large to stay up in the air, they fall to Earth as rain, snow, sleet, or hail. Some of this **precipitation** (pree•sip•uh•TAY•shuhn) collects in lakes, rivers, and other bodies of water. Some precipitation soaks into the ground to become groundwater. Some falls directly back into the ocean.

What Is in Ocean Water

Ocean water is a mixture of water and many dissolved solids. Most of these solids are salts. Sodium chloride is the most common salt in ocean water. You probably know this substance by another name—table salt.

Where do you think the salts and other solids in the ocean come from? Most of the salts and other substances in the ocean come from the land. As rivers, streams, and runoff flow over the land, they slowly break down the rocks that make it up. Over time, flowing water carries substances from the rocks to the ocean.

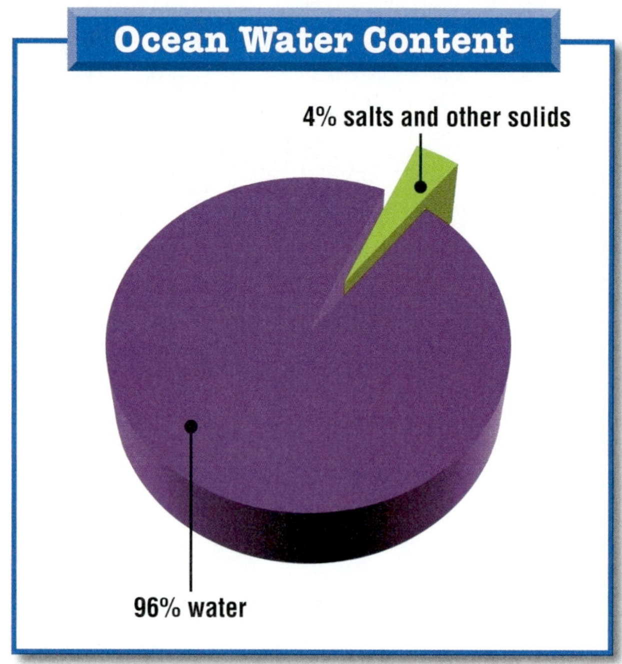

▲ Ocean water is made up of almost the same substances everywhere on Earth. Ocean water is about 96 percent water and 4 percent salts and other dissolved solids.

Dead Sea

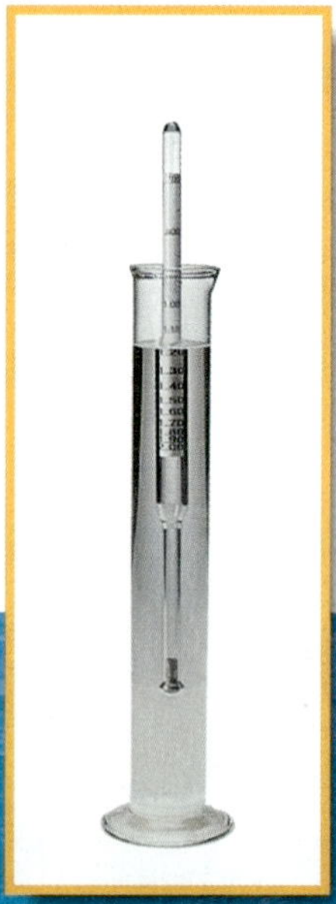

Ocean

Salts and other dissolved substances in the ocean make ocean water denser than fresh water. So it is easier to float in water that is saltier. Look closely at these swimmers and the instruments shown next to them. One swimmer is in the Dead Sea, a body of water that is much saltier than the ocean, and the other is in the ocean. Which water is denser? How can you tell?

Near places where rivers empty into the ocean, the ocean water is less salty than it is farther from the shore. This is because the fresh water mixes with the salt water. Ocean water is a little saltier near the equator, where it is hot and water evaporates faster. And ocean water is a little less salty near the North and South Poles, where it is colder and water evaporates more slowly.

✔ **What is in ocean water?**

Summary

The waters of the ocean provide fresh water for Earth through the water cycle. As water moves through this cycle, it changes from a liquid to a gas and back to a liquid through the processes of evaporation and condensation. The water returns to Earth as precipitation. Sodium chloride is the most common salt in the ocean. The salts and other substances dissolved in ocean water make it denser than fresh water.

Review

1. What is the water cycle?
2. Explain how water changes from a liquid to a gas and back to a liquid in the water cycle.
3. What factors affect the saltiness or density of ocean water?
4. **Critical Thinking** How could you make salt water denser?
5. **Test Prep** Which of these processes occurs when a gas changes to a liquid?
   - A evaporation
   - B condensation
   - C precipitation
   - D salinity

LINKS

MATH LINK

Collect, Organize, and Display Data Use library reference materials to find out more about the amounts of fresh water and salt water on Earth. Draw a large square on a sheet of paper, and divide it into fourths. Color the squares to show the amounts of land and ocean. Stack pennies or checkers on the squares to stand for the amounts of fresh water and salt water.

WRITING LINK

Narrative Writing—Story Suppose you are sailing alone around the world. For your teacher, write down some of your thoughts that describe the ocean and what it is like to have nothing but water all around you.

SOCIAL STUDIES LINK

El Niño Find out what El Niño is. Locate on a world map the places where this condition occurs. Write a report that explains what causes this situation and how it affected weather and crops around the world in 1998.

TECHNOLOGY LINK

Learn more about Earth's water systems by visiting the National Air and Space Museum Internet site.
www.si.edu/harcourt/science

Smithsonian Institution®

LESSON 2

What Are the Motions of Oceans?

In this lesson, you can . . .

INVESTIGATE water currents.

LEARN ABOUT the ways ocean water moves.

LINK to math, writing, social studies, and technology.

Water Currents

Activity Purpose If you've ever gone swimming in the ocean, you've probably felt waves crash against your body. You may also have felt water moving against you below the surface. This movement below the water's surface is a *current.* In this investigation you'll **make a model** and **infer** one way currents form.

Materials

- clear, medium-sized bowl
- warm tap water
- colored ice cube
- clock

Activity Procedure

1. Put the bowl on a flat surface. Carefully fill the bowl three-quarters full of warm tap water.
2. Let the water stand undisturbed for 10 minutes.

◄ Ocean water moves in many ways. Both the rising water and the waves are washing away this sand castle.

3. Without stirring the warm water or making a splash, gently place the colored ice cube in the middle of the bowl. (Picture A)

4. **Observe** for 10 minutes what happens as the ice cube melts. Every 2 minutes, make a simple drawing of the bowl to **record** your observations.

Picture A

Draw Conclusions

1. Describe what you **observed** as the ice cube melted in the bowl of warm water.
2. In your **model**, what does the bowl of water stand for? What does the ice cube stand for?
3. Since the liquid in the bowl and the ice cube were both water, what can you **infer** about the cause of what happened in the bowl?
4. **Scientists at Work** In Chapter 1, you learned that cold air is denser than warm air. The same is true for water. Using this information and what you **observed** in the investigation, explain one way ocean currents form.

Investigate Further Mix up two batches of salt water. Use twice as much salt in one batch as in the other. Use the water to **model** another kind of ocean current. Fill a clear bowl three-fourths full with the less salty water. Add a few drops of food coloring to the saltier water. Along the side of the bowl, slowly pour the colored, saltier water into the clear, less salty water. Describe your **observations. Make a hypothesis** to explain what you observed. What **prediction** can you make based on the hypothesis? **Plan and conduct a simple investigation** to test your hypothesis.

Process Skill Tip

People make models to help them **observe** things in nature that are too small, too big, or too hard to see or understand. By observing a model, you can infer how things work.

Ocean Movements

FIND OUT

- about ocean waves and currents
- what causes tides

VOCABULARY

wave
storm surge
tide
surface current
deep ocean current

Waves

If you've ever been in the path of a wave in the ocean, in a lake, or in a wave pool, you know that even big waves don't move you either forward or back. You bob up and down, but you're still in about the same place after the wave passes. This is because a **wave** is the up-and-down movement of the water particles that make it up.

Water waves are caused by the wind. As wind blows over the water's surface, it pulls on the water particles. This causes small bumps, or ripples, of water to form. As the wind continues to blow, the ripples keep growing. Over time they become waves.

The height of a wave depends on three things: the strength of the wind, the amount of time the wind blows, and the size of the area over which the wind blows. Strong, gusty winds blowing

Waves drop and take away bits of rock and sand grains from a beach as they break on a rocky shore.

Waves like these are caused by the wind. Waves break, or give up their energy, as they move onto the shore.

▲ **Waves erode a shore as the water carries sediment back toward the sea. What might happen to these houses if erosion continues?**

over an area of many square kilometers can cause a group of very large waves called a **storm surge** to form. Storm surges often occur during hurricanes and can cause a lot of damage along a shore.

Waves change the shore in different ways. When waves break on a beach, water carries sand and other sediments as it flows back into the ocean. This carrying away of sediments is called *erosion.* Erosion along a shore causes beaches to become smaller. As waves give up their energy, they also deposit, or drop, sediments. This process is called *deposition* (dep•uh•ZISH•uhn). When waves deposit sediments near shore, a beach gets bigger.

The photograph at the bottom of this page shows a harbor during a hurricane. You probably know that a hurricane is a severe storm that has strong winds and a lot of rain. Storm surges during hurricanes cause erosion and deposition along a shore. Whole beaches can be washed away.

✓ **How do water waves change a shoreline?**

This is a harbor during a hurricane. Storm surges during hurricanes can be as high as 10 meters (more than 30 ft). ▼

Tides

If you watched a beach for 12 hours, you would probably notice that the waves don't always reach the same place. This is because of another type of ocean water motion called tides. **Tides** are the daily change in the local water level of the ocean.

At *high tide* much of the beach is covered by water. At *low tide* waves break farther away from shore. Less of the beach is under water. Every day most shorelines have two high tides and two low tides. High tide and low tide are usually a little more than six hours apart.

Tides are caused by gravity. *Gravity* is a force that causes all objects to be pulled toward all other objects. The force of gravity between two objects depends on two things: the sizes of the objects and the distance between them. Big objects have a greater pull than small objects. Objects that are closer together have a greater pull on one another than objects farther apart do.

Even though the moon is much smaller than the sun, the pull of the moon's gravity on Earth is the main cause of ocean tides. This is because the moon is much closer to Earth than the sun is.

▲ This photograph shows low tide in a harbor on the Bay of Fundy in Nova Scotia, Canada.

▲ This is the same harbor on the Bay of Fundy during high tide. Compare the positions of the ships with their positions in the photograph above.

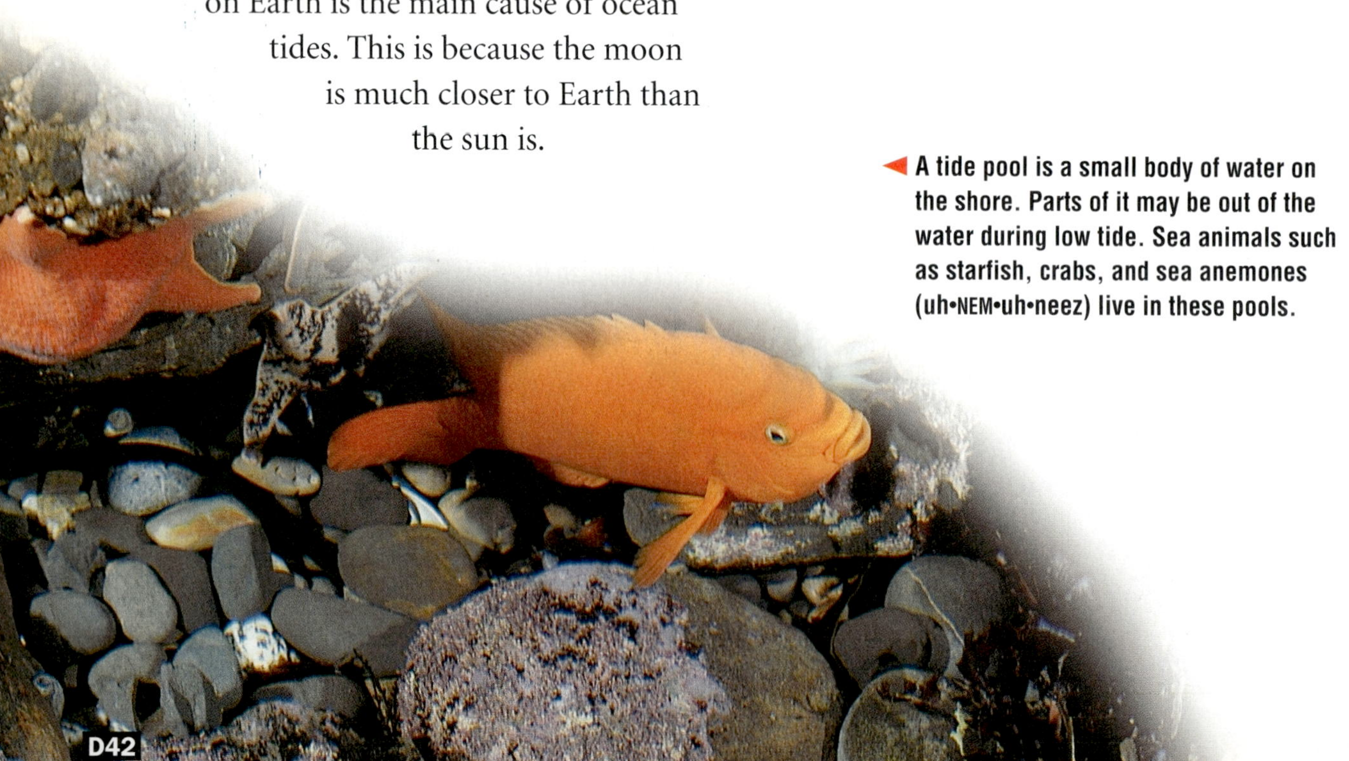

◄ A tide pool is a small body of water on the shore. Parts of it may be out of the water during low tide. Sea animals such as starfish, crabs, and sea anemones (uh•NEM•uh•neez) live in these pools.

The moon pulls on everything on Earth. As the moon pulls on ocean water, the water forms a bulge that always faces the moon. Another water bulge forms on the side of Earth farthest from the moon, where the moon's pull is weakest. As Earth rotates, the bulges stay in the same places. High tide occurs as a point on Earth moves through a bulge. The water level rises on the shore. Low tide occurs as a point on Earth moves between the bulges. The water level on the shore gets lower.

At certain times each month, tides are very high or very low. Read The Inside Story to find out why.

✔ **What are tides?**

THE INSIDE STORY

The Moon and Tides

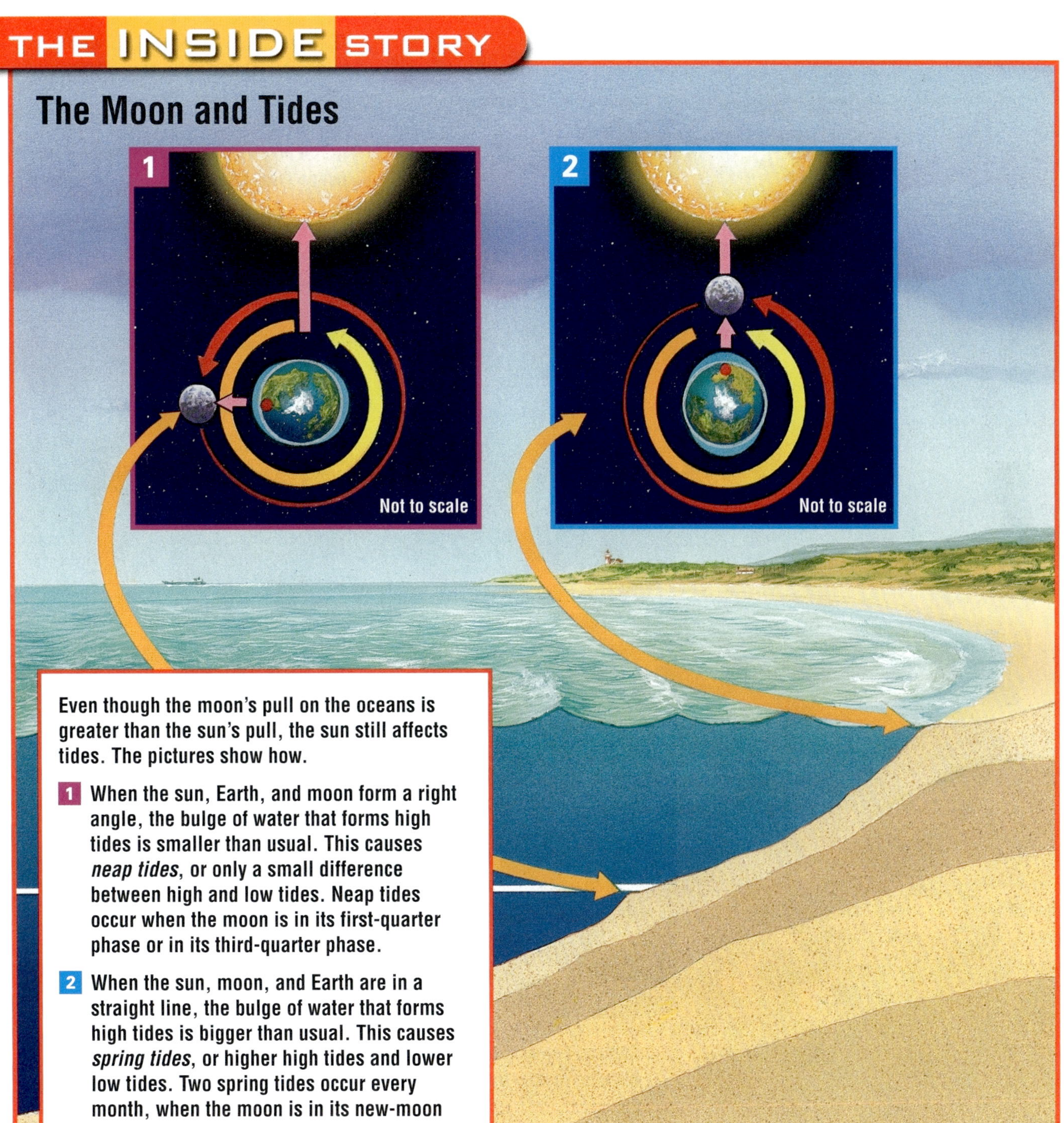

Even though the moon's pull on the oceans is greater than the sun's pull, the sun still affects tides. The pictures show how.

1 When the sun, Earth, and moon form a right angle, the bulge of water that forms high tides is smaller than usual. This causes *neap tides*, or only a small difference between high and low tides. Neap tides occur when the moon is in its first-quarter phase or in its third-quarter phase.

2 When the sun, moon, and Earth are in a straight line, the bulge of water that forms high tides is bigger than usual. This causes *spring tides*, or higher high tides and lower low tides. Two spring tides occur every month, when the moon is in its new-moon and full-moon phases.

Currents

Currents are rivers of water that flow in the ocean. A **surface current** forms when steady winds blow over the surface of the ocean. In the Northern Hemisphere, surface currents flow in a clockwise direction. In the Southern Hemisphere, surface currents flow in a counterclockwise direction.

Deep ocean currents form because of density differences in ocean water. You made a model of deep ocean currents in the investigation. The density of ocean water depends on two things—the amount of salt in the water and the temperature of the water. The more salt there is in water, the denser it will be. Cold ocean water also is

Scientists have mapped the paths of surface currents and deep ocean currents in the oceans.

Cold water flows underneath warm water, forming a deep ocean current. The cold water can flow for great distances along the ocean floor. The green arrows on the globe show deep ocean currents.

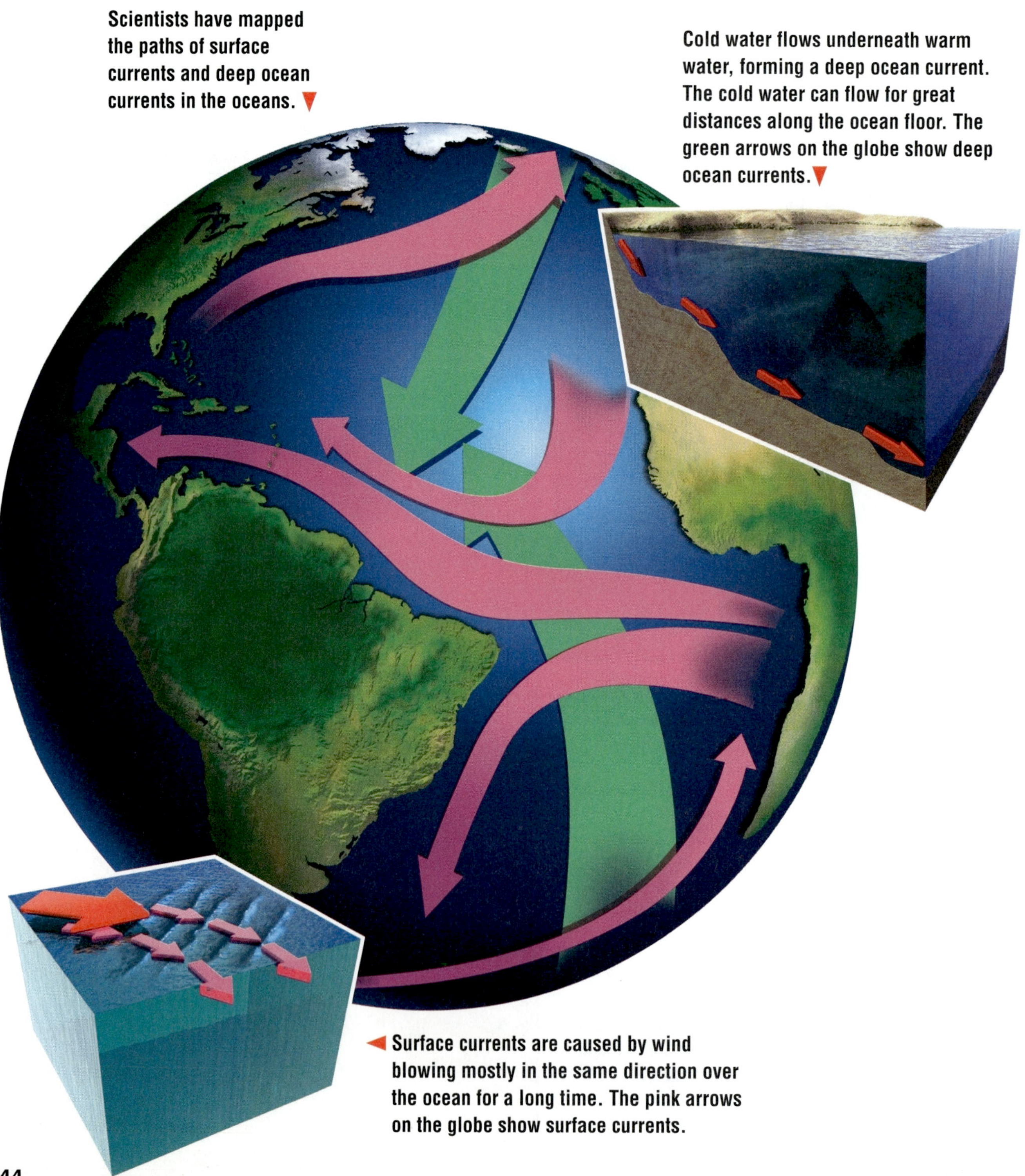

Surface currents are caused by wind blowing mostly in the same direction over the ocean for a long time. The pink arrows on the globe show surface currents.

denser than warm ocean water. Deep ocean currents form when dense, cold water meets less dense water. The denser water flows under the less dense water, forcing the less dense water to rise.

✔ **What causes deep ocean currents?**

Summary

Ocean waves form as wind blows over the water's surface. Strong, gusty winds over a large area can cause a large wave called a storm surge. Tides are the rise and fall of ocean water caused by the pull of gravity between Earth, the moon, and the sun. Surface currents are caused by the blowing of steady winds. Deep ocean currents are caused by differences in saltiness or water temperature.

Review

1. How does an object floating in water move as a wave passes it?
2. What causes tides?
3. Why does the moon have a greater effect on tides than the sun has?
4. **Critical Thinking** Compare and contrast surface currents in the Northern and Southern Hemispheres.
5. **Test Prep** Which is caused by density differences?
   A surface currents
   B ocean waves
   C deep ocean currents
   D ocean tides

LINKS

MATH LINK

Find an Average Scientists estimate that sea level rose 10 to 15 centimeters between the years 1900 and 2000. They estimate that it will rise another 50 centimeters before 2100. What will the average yearly rate of sea level rise be for the years 2000 to 2100?

WRITING LINK

Informative Writing—Description Suppose you are a creature that lives in a tide pool. For a younger student, write a short description that explains how your life changes when the tide comes in and goes out.

SOCIAL STUDIES LINK

Trade Routes Find out about the triangle trade route that existed in colonial times in the United States. Mark this route on a copy of a world map, and explain how ocean currents and wind patterns made this route possible.

TECHNOLOGY LINK

Learn more about new ways to explore deep in the ocean by viewing *Robot Submarine* on the **Harcourt Science Newsroom Video.**

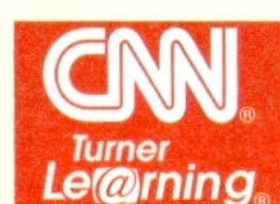

LESSON 3

What Is the Ocean Floor Like?

In this lesson, you can . . .

INVESTIGATE the ocean floor.

LEARN ABOUT features of the ocean floor.

LINK to math, writing, social studies, and technology.

Model the Ocean Floor

Activity Purpose The sonar ship in the picture uses an echo sounder to gather data about the ocean floor. This device sends sounds to the ocean floor. The sounds echo, or are reflected, back to the ship. The time for the signal to travel and return is measured. Using these times, a computer can calculate and display ocean floor depths. In this activity you will **use numbers** obtained by echo sounders to make a model of ocean floor features.

Materials

- grid paper
- shoe box
- clay

Activity Procedure

1. Set up a graph as shown at the top of page D47. Label the horizontal axis *Distance East of New Jersey (km)*. Label the vertical axis *Water Depth (m)*.
2. Look closely at the graph. Note that the top horizontal mark is labeled *0*. This mark stands for the surface of the ocean. The numbers beneath stand for depths below sea level.
3. Plot the chart data on the graph.
4. Connect the points on your graph. You have now made a profile of the Atlantic Ocean floor.
5. **Analyze** your graph. Determine how the ocean floor changes as you move eastward from New Jersey.

This is a sonar survey ship. Using a device called an echo sounder, the ship gathers data about ocean floor features. ▼

Distance (km)	Depth (m)
0	0
500	3500
2000	5600
3000	4300
4000	0
4300	3050
5000	5000
5650	0

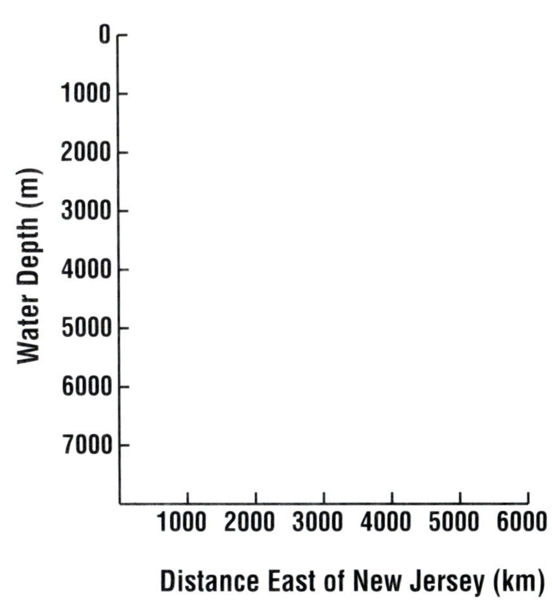

6 Use clay to **make a model** of your profile. Label the narrow end of the shoe box *Coastline*. Label the opposite end *6000 km*. Put the clay inside the box. Vary the height of the clay to model the changing depth of the ocean floor. (Picture A)

Picture A

Draw Conclusions

1. Describe what you **observed** on the graph.
2. What does the clay in your **model** represent?
3. Based on your model, what can you infer about the ocean floor?
4. **Scientists at Work** Sound waves travel through ocean water at about 1525 meters per second. Suppose it takes two seconds for an echo to return. The sound takes one second to reach the ocean floor and one second to return, so the distance to the ocean floor is about 1525 meters. **Using the numbers** from the chart, at what distance from New Jersey would it take a sound wave four seconds to return to a ship?

Investigate Further Use numbers to make a chart of heights to profile one section of your classroom. Your chart should resemble the one shown in this activity. Exchange charts with a classmate and try to identify the exact location shown by the map.

Process Skill Tip

Scientists often **use numbers** to make graphs and models. The numbers help them accurately show how objects in the real world look or move.

The Ocean Floor

FIND OUT

- about some ocean floor features
- how ocean floor forms

VOCABULARY

shore zone
continental shelf
abyssal plain
trench
mid-ocean ridge

Underwater Geography

What would you see if you somehow drained the water from the world's oceans? Many people think you would find a big, flat plain. In some places the ocean floor does look like that. But the ocean floor also holds the world's longest mountain range. This chain of mountains is four times longer than the Andes, Rockies, and Himalayas combined. Earth's tallest mountain is on the ocean floor. It is more than 1220 meters (about 4000 ft) taller than the tallest mountain on dry land. In some places, the ocean floor drops thousands of meters to form deep trenches, or canyons. One place in the Pacific Ocean is almost seven times as deep as the Grand Canyon.

In the investigation you made a profile of the floor of the Atlantic Ocean. The profile of the Atlantic Ocean shown on these pages was made in a similar way but with many more data points. Study the profile to learn more about the features of the ocean floor.

✔ **How is the ocean floor like Earth's surface?**

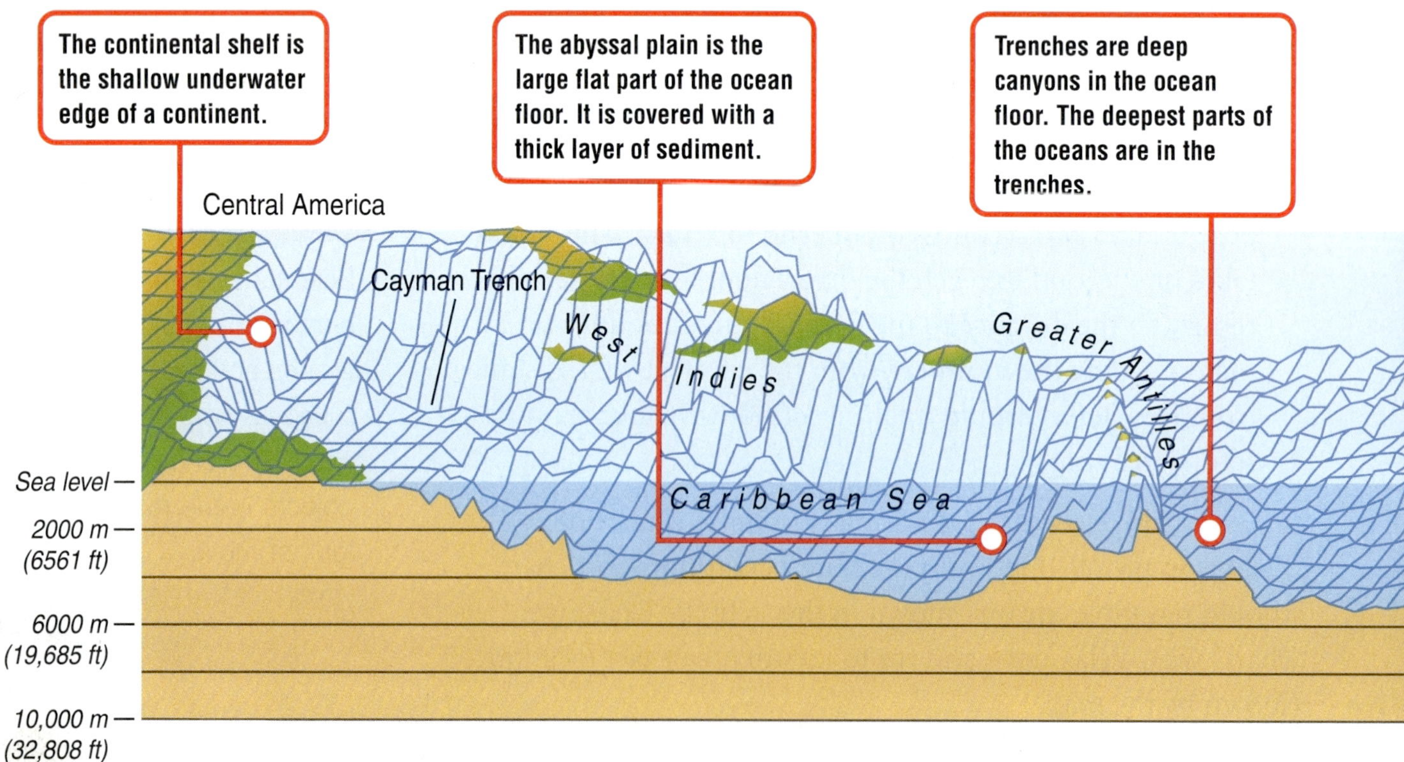

Near the Shore

The **shore zone** is the place where land and ocean meet. The ocean floor of the shore zone is called the **continental shelf**. Here the continent slopes gently down beneath the water's surface. At the edge of a continental shelf, the ocean floor drops away rapidly.

The shore zone is an area rich in resources. Fish caught in the shore zone are an important food source. Oils and fertilizers are made from fish. Many types of salts are obtained from water in the shore zone.

✓ What are some products of the shore zone?

Some parts of the shore zone contain oil deposits. Offshore rigs drill holes in the continental shelf to bring this valuable resource to Earth's surface.

◄ Rocks and sand obtained from the shore zone are used to make roads and buildings.

◄ Dark masses of minerals form along the ocean floor. In the future these nodules may be mined for the metals they contain—manganese, copper, nickel, and cobalt.

This picture shows how the continental shelf slopes down under the ocean. ►

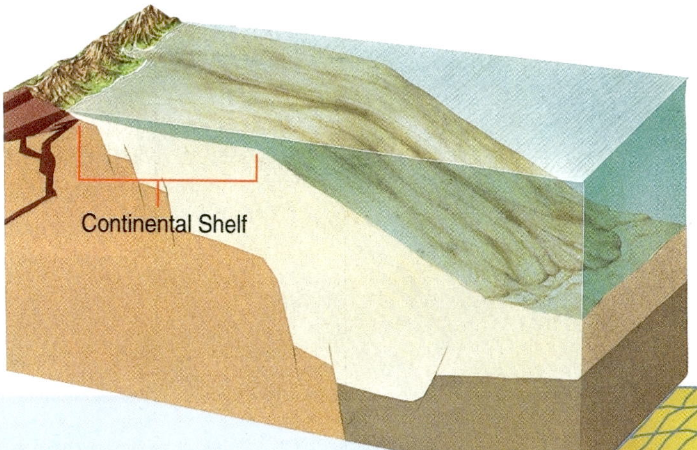

The mid-ocean ridge is the world's longest mountain range. It extends across 65,000 km (about 40,000 mi) of the ocean floor.

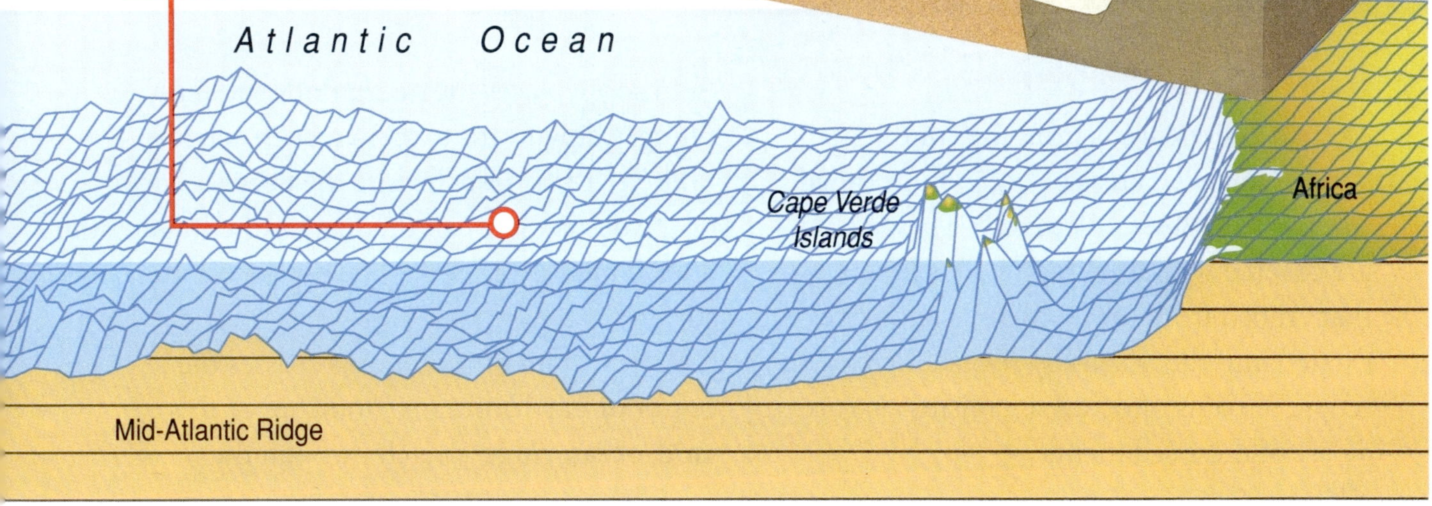

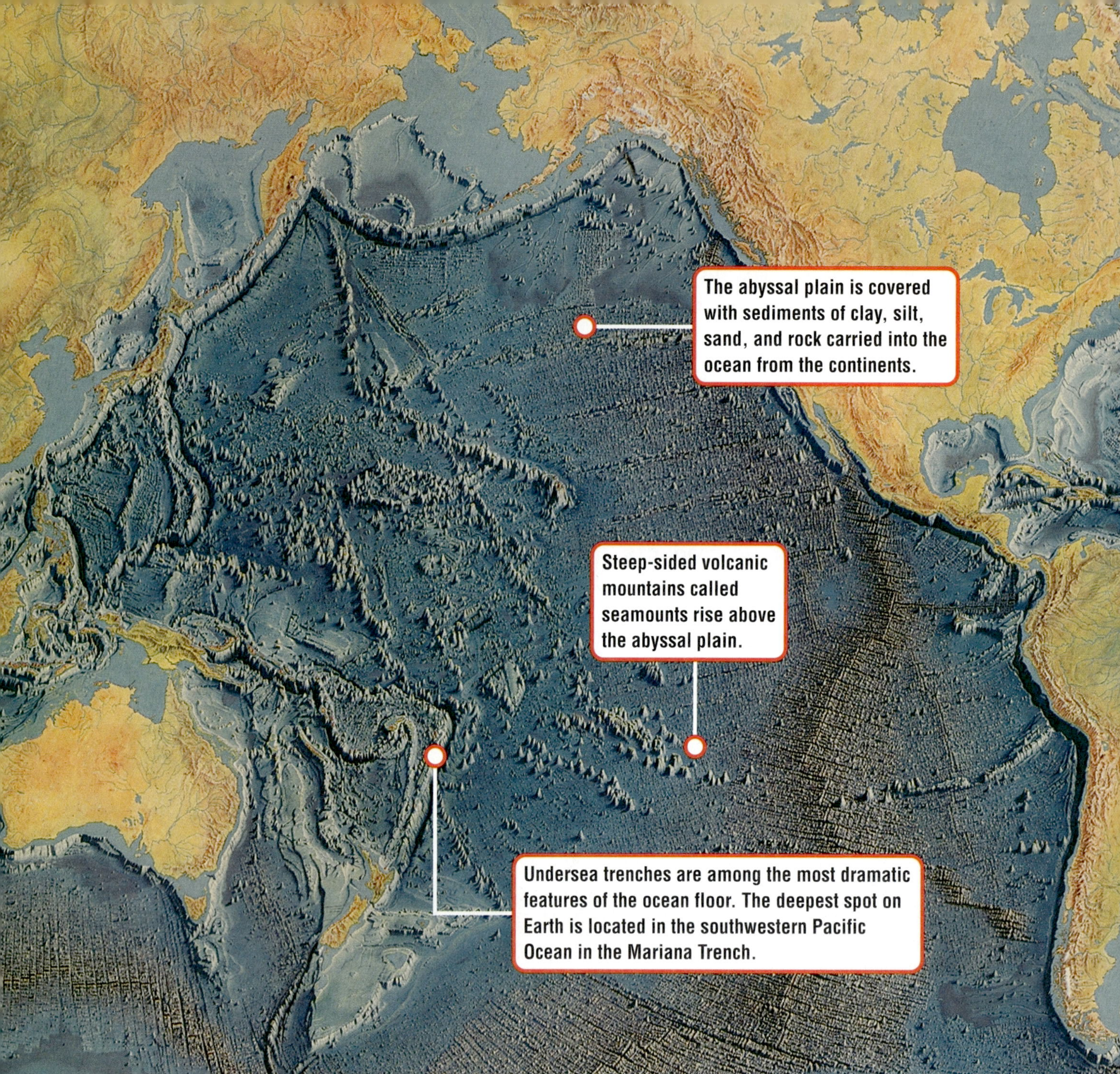

Mid-Ocean Features

A steep slope leads from the continental shelf to huge flat areas of ocean floor called **abyssal** (uh•BIHS•uhl) **plains**. These plains are covered with thick layers of sediment that eroded from the continents and were carried into the oceans by rivers. Microscopic remains of ocean organisms also are a part of these sediments.

In some places the ocean floor suddenly plunges downward to form a deep, narrow **trench**. These trenches are the deepest parts of the ocean. Some of them are more than 10,000 meters (32,800 ft) below the ocean surface.

A vast chain of mountains crosses the center of Earth's oceans. Branches of this **mid-ocean ridge** stretch through the

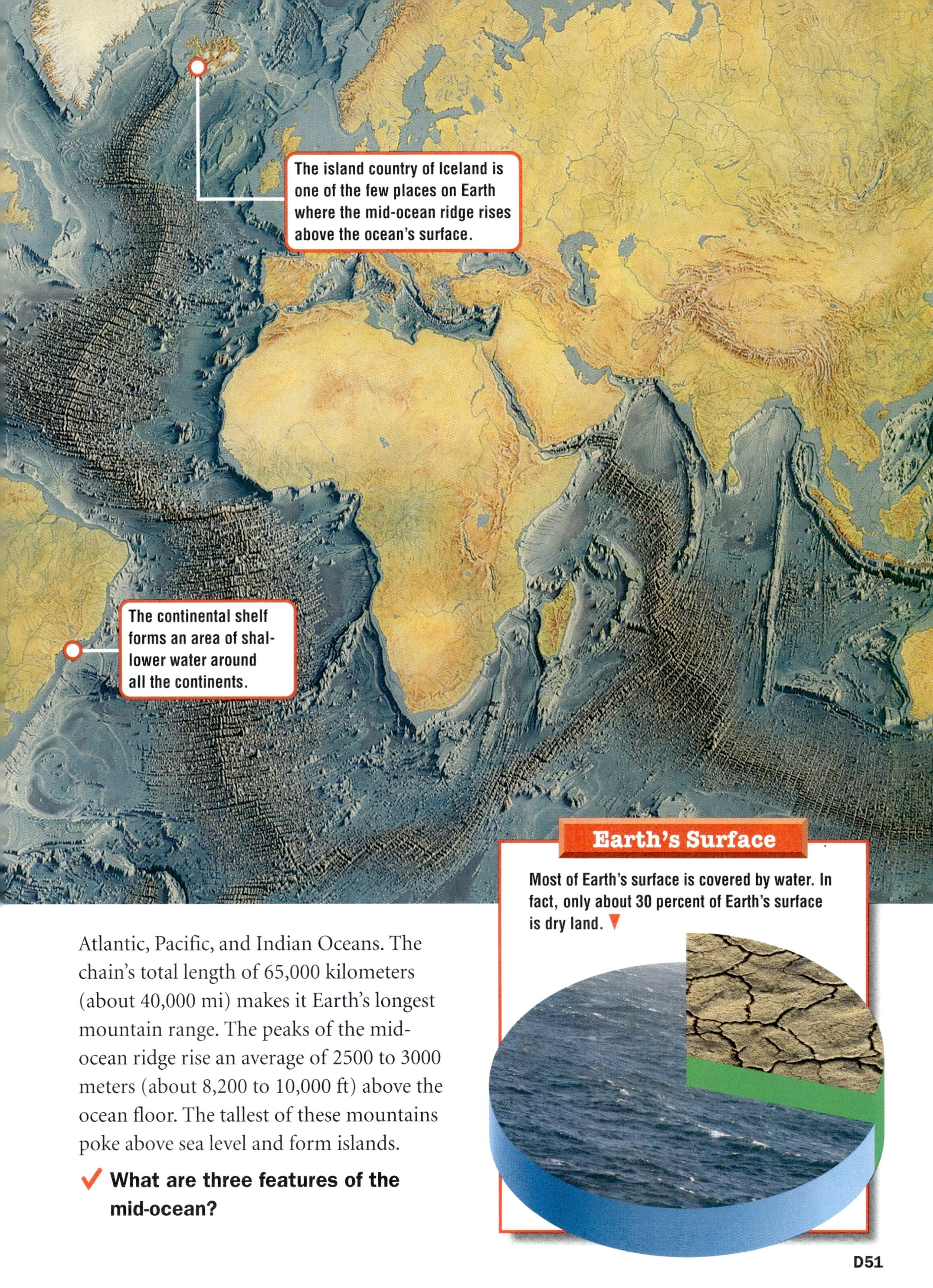

Earth's Surface

Most of Earth's surface is covered by water. In fact, only about 30 percent of Earth's surface is dry land. ▼

Atlantic, Pacific, and Indian Oceans. The chain's total length of 65,000 kilometers (about 40,000 mi) makes it Earth's longest mountain range. The peaks of the mid-ocean ridge rise an average of 2500 to 3000 meters (about 8,200 to 10,000 ft) above the ocean floor. The tallest of these mountains poke above sea level and form islands.

✔ **What are three features of the mid-ocean?**

The ocean floor is continually recycled through seafloor spreading. New crust forms in places where two ocean plates move apart. At other locations old crust is destroyed in ocean trenches. Scientists compare this movement of ocean floor to the motion of a conveyor belt. In both processes matter moves in a repeating pattern.

Basalt, a dark, heavy rock, forms from the magma that moves up to fill the crack formed by the plates moving apart. These ridges of basalt are what pile up to form mid-ocean ridges.

Changing Features

You may recall from Unit C, Chapter 1, that movement of Earth's plates produces many features of the ocean floor. Some underwater volcanoes form when a plate moves over a hot spot in Earth's mantle. An ocean trench forms when one plate slides beneath another.

Plate movement also produces the mid-ocean ridges and new ocean floor. At the ridges two plates are moving very slowly away from each other. As the plates move apart, magma from the mantle moves up between them. This cooled magma forms new ocean-floor crust. The new crust replaces older crust pushed deep into Earth at a trench. Sometimes the new crust collects to form a mountain.

✔ **What causes the formation of new ocean floor?**

Summary

Many of the features found on dry land also are found on the floors of Earth's oceans. In the shore zone, a continent slopes beneath the water's surface. This area of the ocean supplies humans with many valuable products. Far away from the shore zone, a long mountain ridge crosses the ocean floor. Other features of the ocean floor include vast abyssal plains and deep ocean trenches. The movement of Earth's plates forms many of these features. Plate movement also causes the formation of new ocean floor and the destruction of old ocean floor.

Review

1. How are the shore zone and continental shelf related?
2. What is unusual about Iceland?
3. How does new ocean floor form?
4. **Critical Thinking** What would happen if old ocean floor were not destroyed in deep trenches?
5. **Test Prep** Which ocean floor feature forms when one plate slides beneath another plate?

   A abyssal plain
   B trench
   C shore zone
   D volcano

LINKS

MATH LINK

Use Mental Math Ten liters of sea water weigh about 101 newtons. How much would a 4,000 liter tank weigh?

WRITING LINK

Informative Writing—Description Suppose you are a passenger in a deep-sea submarine. Write a letter to a friend describing your trip from the shore zone to the mid-ocean ridge.

SOCIAL STUDIES LINK

Ocean Profile Choose one of Earth's oceans. Research the unique features found on the floor of this ocean. Display your findings as a profile of that ocean.

PHYSICAL EDUCATION LINK

Scuba Diving Many people enjoy exploring the undersea world through the sport of scuba diving. Research the equipment and training needed to enjoy this activity. Share your findings in a pamphlet titled *Safe Scuba Diving Tips*.

TECHNOLOGY LINK

Learn more about Earth's ocean floor by visiting this Internet site.

www.scilinks.org/harcourt

Deep Flight II

Deep Flight II is a submersible, or sub, that is being developed to explore the ocean depths. It will carry two pilots and will go to the deepest part of the ocean, the Mariana Trench.

Exploring the Deep Ocean

Only one mission with people has ever gone to the Mariana Trench. This trench in the Pacific Ocean is about 11,000 meters (about 7 mi) deep. From that depth, it would take more than 25 Empire State Buildings stacked on top of each other to reach the surface. In 1960 two men in a bathyscaph (BATH•uh•skaf) called the *Trieste* (tree•EST) went down for 20 minutes. The ship had none of the video cameras or computers we have today. And the *Trieste* could only go straight down and come straight back up.

Graham Hawkes, an engineer, and his business partner, Sylvia Earle, have been working to make deep-ocean subs, such as *Deep Flight II*. Scientists want better subs to help them investigate ocean-floor geology as well as deep-ocean plants and animals. Subs, however, can't meet all research needs. Robot, or remote-controlled, subs are often better for dangerous or long trips.

Built for Speed and Comfort

The time needed to go so deep is a problem, so Hawkes is designing a craft that "flies" through water like an airplane. It should reach the ocean floor in 90 minutes.

Another problem is the high pressure. The sub must support the weight of a column of water 11,275 meters high, so the hull of *Deep Flight II* is being built of strong, new ceramic materials. These materials are lighter than steel and won't break under high pressure. The ship will also protect its crew from the near-freezing cold of the ocean water.

Hawkes is using shapes from nature for the design of *Deep Flight II*. Its smooth body and wings look like parts of dolphins, whales, and birds, as well as aircraft. It can skim forward below the ocean's surface or dive down into the depths. It can even do spins and rolls like a stunt airplane.

If the first voyage of *Deep Flight II* is a success, Hawkes and Earle may soon be designing and building more deep-sea subs.

Graham Hawkes in a museum model of *Deep Flight I*

THINK ABOUT IT

1. How will the pilot of *Deep Flight II* be like an astronaut landing on the moon?
2. Why do you think most of the ocean remains unexplored?

CAREERS
SCUBA SUPPORT CREW

What They Do People working on a scuba support crew prepare equipment for diving. They make sure that everything needed for a dive *works*. The scuba support crew stays on land or in a boat while the diver goes under water. The crew communicates with the diver and takes care of any emergencies.

Education and Training
Scuba support crew members must have emergency medical training and a Divemaster certificate from the Professional Association of Diving Instructors.

WEB LINK
For Science and Technology updates, visit the Harcourt Internet site.
www.harcourtschool.com

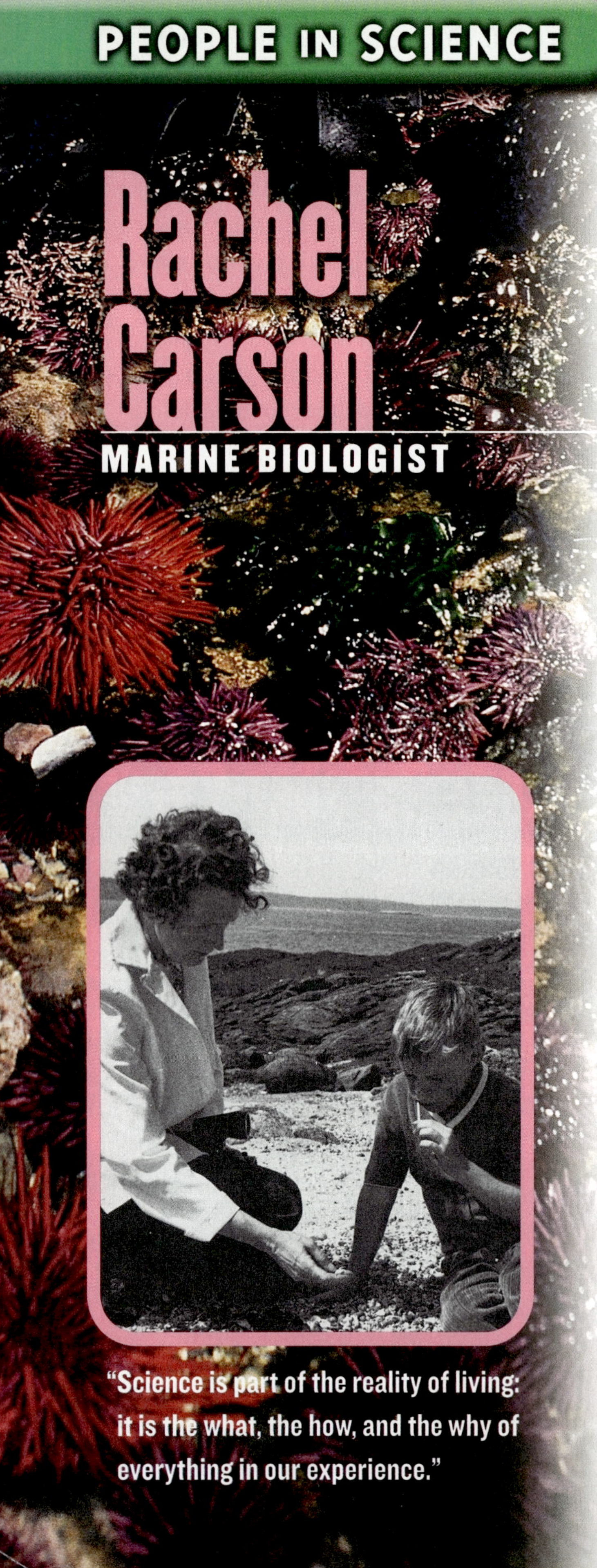

Rachel Carson

MARINE BIOLOGIST

"Science is part of the reality of living: it is the what, the how, and the why of everything in our experience."

Although Rachel Carson did not actually visit the ocean until she was 22 years old and a college graduate, she had been fascinated by it all her life.

Everything in nature thrilled Carson—flowers and birds, trees and rivers, animals and insects. She surprised many people at her college by changing from an English major to a biology major. At the time, science was seen as a career for men. No one expected her to do well.

Carson used both biology and writing in her work. She taught college classes after she graduated. Another of her early jobs was with the U.S. Bureau of Fisheries, writing scripts for radio broadcasts about life under the sea. She eventually moved up in the Bureau to become editor-in-chief of publications.

Carson was encouraged to write articles, which eventually were printed in one book, *Under the Sea-Wind*. Carson wrote other books about the sea, including *The Sea Around Us*, which became a best-seller.

Carson is best known for her book *Silent Spring*. She began it after a friend who had a bird sanctuary wrote a letter. Her friend wrote Carson that pesticides had killed all the birds at the sanctuary. It took Carson several years to collect information and write the book. She closely studied the dangers of DDT, a chemical used to kill insects. *Silent Spring* tells what a spring would be like without new life.

THINK ABOUT IT

1. Why was it important that Carson spend so much time collecting information for *Silent Spring*?
2. Why might it have been important that Carson had already published several books before writing *Silent Spring*?

ACTIVITIES FOR HOME OR SCHOOL

MEASURING DENSITY

What are the relative densities of different solutions?

Materials

- unsharpened pencil with eraser
- tall, narrow glass or jar
- permanent marker
- safety goggles
- thumbtack
- water
- salt
- spoon

Procedure

1. **CAUTION** **Put on your goggles.** Carefully push the thumbtack completely into the eraser.
2. Fill the glass three-fourths full of tap water. Place the pencil, eraser end first, into the water. When it floats up on its own, grab the pencil at the point where it meets the surface of the water. Use the permanent marker to mark this point.
3. Add four spoonfuls of salt to the water. Stir until the salt dissolves. Place the pencil in the water, and mark it as you did in Step 2.

Draw Conclusions

Explain your observations in Steps 2 and 3.

MAKING WAVES

How does a wave move?

Materials

- safety goggles
- heavy washer
- 2-m length of rope

Procedure

1. **CAUTION** **Put on your goggles.** String the washer onto the rope. It should fit loosely.

2. Work with a partner. Each of you should hold one end of the rope. It should hang loosely with the washer in the center of the rope. The rope should not be stretched between you and your partner.
3. Shake one end of the rope by moving your arm while your partner holds the other end still. Observe the movement of the washer on the rope. Compare this movement to the movement of objects floating in ocean water.

Draw Conclusions

How did the washer move? How is the movement of the rope like the movement of a wave?

CHAPTER 2 Review and Test Preparation

Vocabulary Review

Use the terms below to complete the sentences. The page numbers in ( ) tell you where to look in the chapter if you need help.

evaporation (D34)
water cycle (D34)
condensation (D34)
precipitation (D35)
wave (D40)
storm surge (D41)
tides (D42)
surface current (D44)
deep ocean current (D44)
shore zone (D49)
continental shelf (D49)
abyssal plain (D50)
trench (D50)
mid-ocean ridge (D50)

1. A large wave caused by strong winds is called a ____.
2. ____ is a process that changes a liquid to a gas.
3. A gas changes to a liquid by a process called ____.
4. Blowing winds form a ____, a river of water in the ocean.
5. Water vapor, clouds, rain, and the ocean are parts of the ____.
6. Neap and spring ____ are examples of changes in ocean level caused by the pull of the sun, moon, and Earth.
7. ____ is any form of water that falls from clouds.
8. A ____ forms when cold, dense ocean water meets and flows beneath warmer ocean waters.
9. The up-and-down motion of water is called a ____.
10. The longest chain of mountains on Earth is the ____.
11. The ____ is the gently sloping part of the ocean floor near the shore.
12. When two of Earth's ocean plates collide, they can form a deep valley called a ____.
13. An ____ is a big, flat area of the ocean floor that is covered with a thick layer of sediments.
14. Many important resources are harvested from the ____ of the oceans.

Connect Concepts

This diagram shows how water moves from the land and oceans to the atmosphere and back again. Label each section of the diagram, and write a title for it.

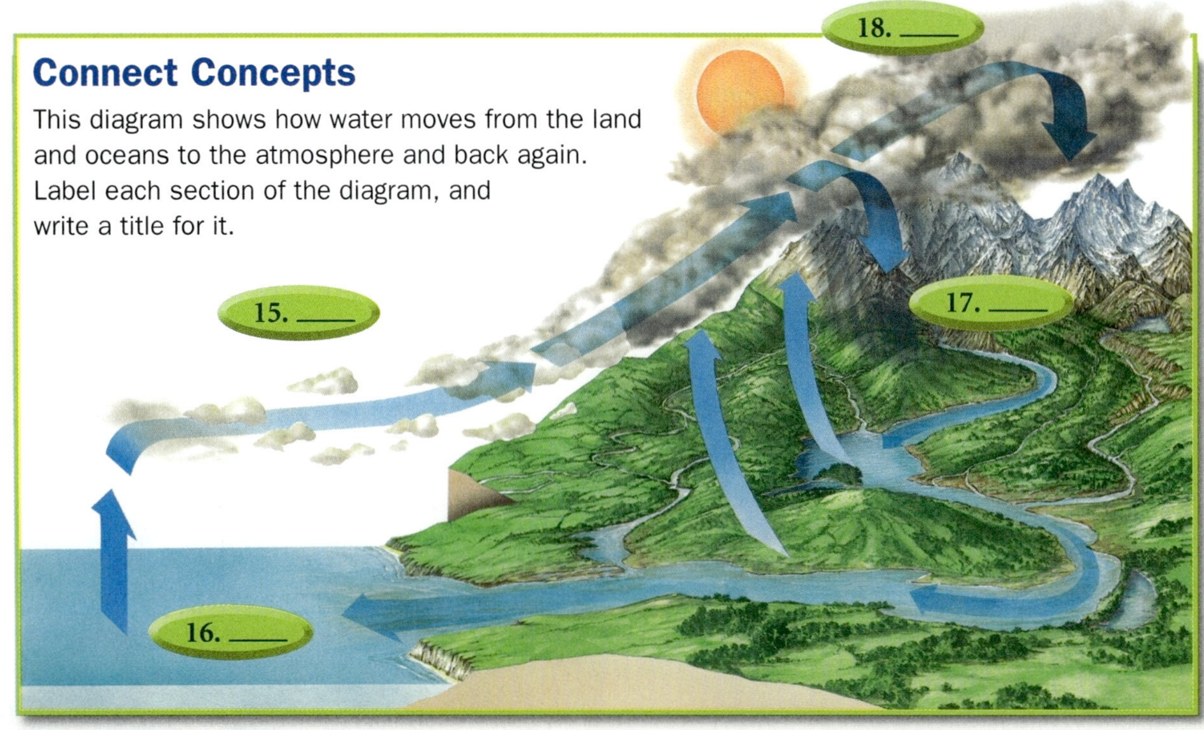

Check Understanding

Write the letter of the best choice.

19. The ____ provides the energy for the water cycle.

A moon C atmosphere
B sun D ocean wave

20. Ocean water is a mixture of water and many —

F gases H liquids
G salts J fluids

21. Wind causes an up-and-down movement of ocean water, called a —

A deep ocean current
B tide
C surface current
D wave

22. The part of the ocean floor nearest the shore is the —

F abyssal plain
G mid-ocean ridge
H continental shelf
J trench

23. The pull of ____ on Earth is the main reason for tides.

A the moon
B winds
C the sun
D currents

Critical Thinking

24. Would it be easier to float in the Great Salt Lake or in Lake Michigan? Explain your answer.

25. What do you think would happen to tides if the moon's gravity had a stronger pull than it does now?

Process Skills Review

26. Suppose there is a heavy rainstorm far out over the ocean. What can you **infer** about the density of the ocean water at the surface in that area right after the storm?

27. You **observe** clouds forming on a bright, sunny day. What can you **infer** is happening in the atmosphere? What may happen later in the day?

28. Sometimes people **use numbers** to compare very large natural objects to an everyday object. Pick an everyday object, estimate its size, and use it to describe one of the ocean floor features on pages D48–D53.

Performance Assessment

Currents

Completely fill a plastic cup with hot water. Add three or four drops of food coloring to the water. Cover the cup with plastic wrap. Hold the wrap in place with a rubber band.

Put the cup inside a bowl. Fill the bowl with cold water until it is almost full. There should be 2–3 cm of water over the cup. Use a pencil to poke a hole in the plastic wrap. Observe what happens. Explain what is happening inside the bowl. How could you make the same thing happen using salt water instead of hot water?

CHAPTER 3

Planets and Other Objects in Space

Vocabulary Preview

satellite
orbit
phase
revolution
axis
rotation
solar system
star
planet
asteroid
comet
inner planet
outer planet
gas giant
telescope
space probe

From Earth you can study objects in space by just stepping outside on a clear night. Most of the objects you will see are stars, which are very, very distant suns. A few of the objects you will see are planets. Some are a little like Earth, and some are amazingly different.

Ulysses orbiting the sun

Fast Fact

The sun is about 150,000,000 kilometers (93 million mi) from the Earth. It would take you about 193 years to travel this distance in a car at highway speed!

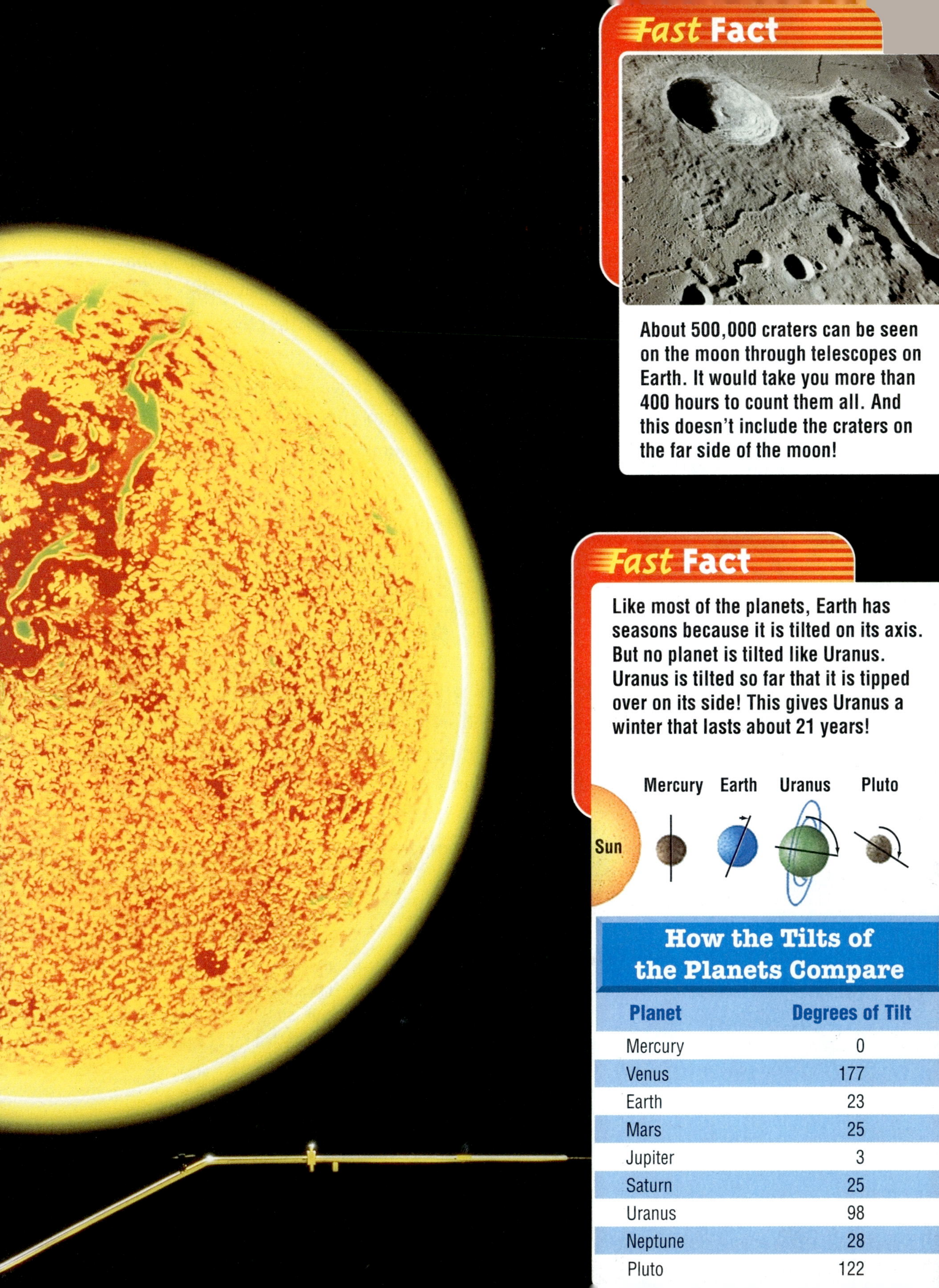

Fast Fact

About 500,000 craters can be seen on the moon through telescopes on Earth. It would take you more than 400 hours to count them all. And this doesn't include the craters on the far side of the moon!

Fast Fact

Like most of the planets, Earth has seasons because it is tilted on its axis. But no planet is tilted like Uranus. Uranus is tilted so far that it is tipped over on its side! This gives Uranus a winter that lasts about 21 years!

How the Tilts of the Planets Compare

Planet	Degrees of Tilt
Mercury	0
Venus	177
Earth	23
Mars	25
Jupiter	3
Saturn	25
Uranus	98
Neptune	28
Pluto	122

LESSON 1

How Do Earth and Its Moon Move?

In this lesson, you can . . .

- INVESTIGATE the movements of Earth and the moon.
- LEARN ABOUT the seasons.
- LINK to math, writing, social studies, and technology.

INVESTIGATE

Relative Size

Activity Purpose The moon is about 384,000 km (238,613 mi) away from Earth. This makes the moon Earth's closest neighbor in space. On the other hand, the sun is a very distant neighbor. It is almost 400 times farther away from Earth than the moon is.

These distances in space affect the way the moon and the sun appear to viewers on Earth. In this investigation you'll **observe** various objects to determine how distance affects image.

Materials

- tennis ball
- basketball
- meterstick

Activity Procedure

1. Work with two other students. Two of you should stand side by side. One of these students should hold a tennis ball. The other should hold a basketball. Have the third student stand 3 to 4 m from the pair and **record** his or her **observations** of the two balls. (Picture A)

◄ Because of its closeness to Earth, the moon is the brightest object in the night sky. The moon is visible because it reflects light from the sun. So, a "moonbeam" is actually a reflected ray of sunshine.

Picture A

Picture B

2. Tell the person holding the basketball to move backward until the basketball appears to be the same size as the tennis ball. Use a meterstick to **measure** the distance between the two students. **Record** this distance. (Picture B)

3. Tell the person holding the basketball to continue to move backward until the basketball appears to be smaller than the tennis ball. Again use the meterstick to **measure** the distance between the two students. **Record** this distance.

Draw Conclusions

1. How did the size of the balls compare when held side by side?
2. At what distance did the balls appear the same size?
3. At what distance did the basketball appear smaller than the tennis ball?
4. The relationship between the sizes of the tennis ball and the basketball is similar to that between the sizes of the moon and the sun. From your **observations**, what can you infer about the size of these neighbors in space when viewed from Earth?
5. **Scientists at Work** How could you make sure that your **observations** of two objects show their actual traits?

Investigate Further Pluto is the farthest planet from the sun. **Hypothesize** how Pluto would appear when viewed from Earth's surface.

Process Skill Tip

You **observe** by using your senses. Scientists gather data by careful observations. They also note the conditions under which the observations were made. In this way, they can make more accurate inferences about what they have observed.

Moon and Earth Orbits

FIND OUT

- how Earth and the moon move
- what causes seasons

VOCABULARY

satellite
orbit
phases
revolution
axis
rotation

Motions of the Moon

The moon is a natural satellite of Earth. A **satellite** is an object that moves around another object in space. The moon moves around Earth in a certain path or **orbit**. It takes a little more than 28 days for the moon to complete its orbit.

When viewed from Earth, the moon appears to be almost the same size as the sun. As you saw in the investigation, these types of observations can be tricky. The moon is actually smaller than Earth, and Earth is much smaller than the sun. The moon looks as large as the sun because it is much closer to Earth.

The moon does not give off its own light. Instead, it reflects light from the sun. Half of the moon faces the sun and so it is lit. As the moon moves through its orbit, different amounts of its lit half can be seen from Earth. That's why the moon seems to have different shapes, or **phases**. The moon's cycle of phases takes just over 28 days to complete.

✓ **What are phases of the moon?**

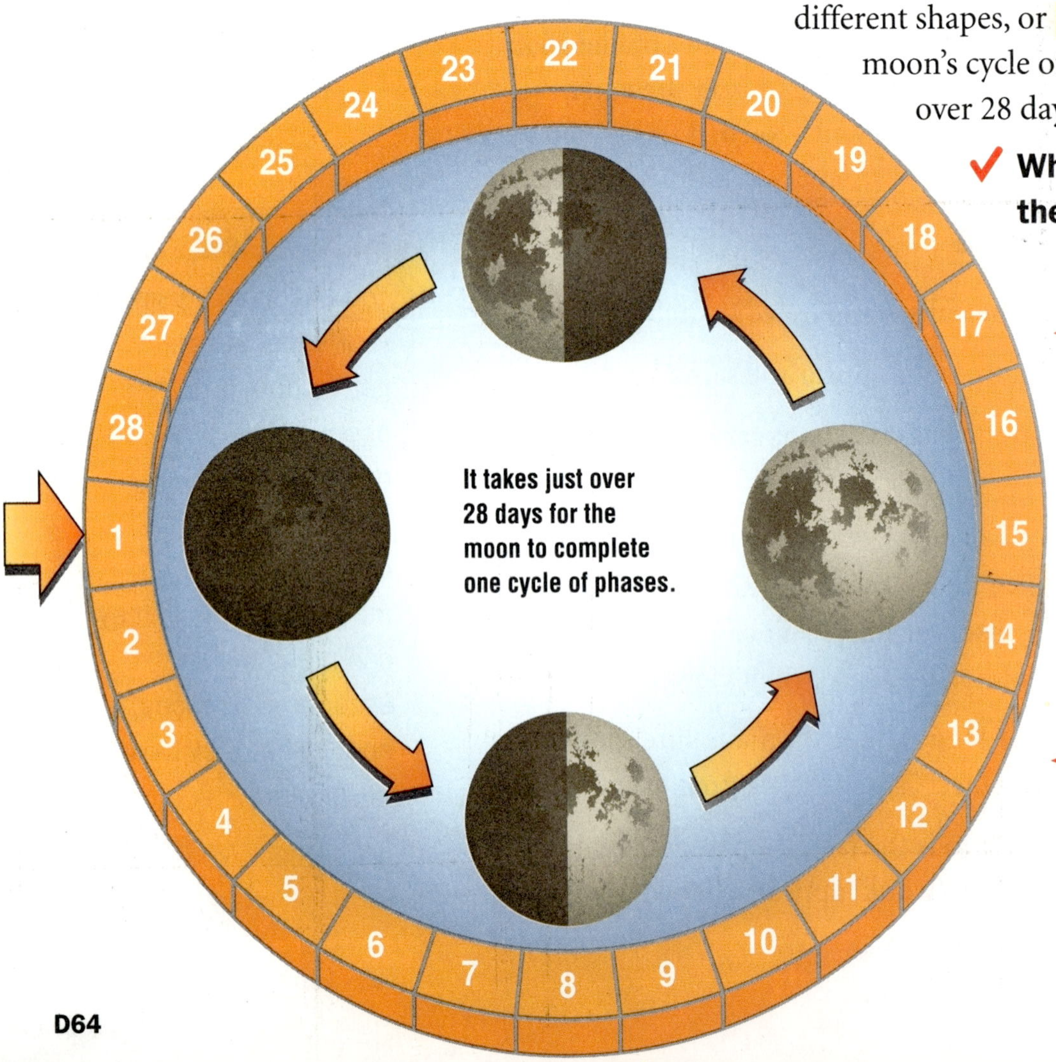

It takes just over 28 days for the moon to complete one cycle of phases.

During the last half of the moon's cycle (days 15 to 28), the amount of the lit side of the moon seen from Earth *wanes*, or decreases. Then the cycle begins again.

During the first half of the moon's cycle (days 1 to 14), the amount of the lit side of the moon seen from Earth *waxes*, or increases.

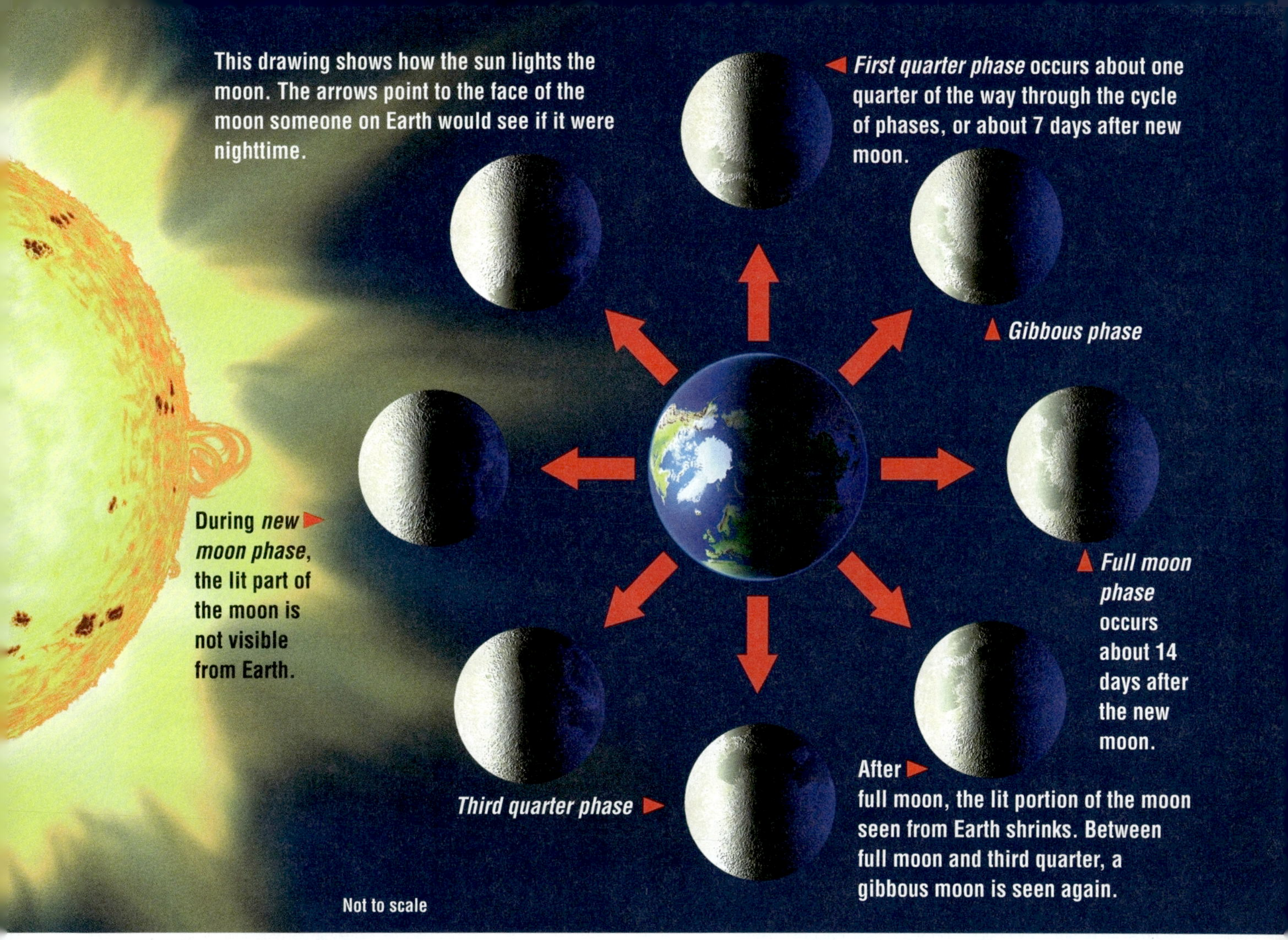

This drawing shows how the sun lights the moon. The arrows point to the face of the moon someone on Earth would see if it were nighttime.
First quarter phase occurs about one quarter of the way through the cycle of phases, or about 7 days after new moon.
Gibbous phase
During new moon phase, the lit part of the moon is not visible from Earth.
Full moon phase occurs about 14 days after the new moon.
After full moon, the lit portion of the moon seen from Earth shrinks. Between full moon and third quarter, a gibbous moon is seen again.
Third quarter phase
Not to scale

Motions of Earth

Just as the moon moves around Earth, Earth moves around the sun. Just as the moon is a satellite of the Earth, Earth is a satellite of the sun. The movement of Earth around the sun is called its **revolution**. It takes Earth one year (365 days) to complete a revolution.

As Earth revolves around the sun, it is also spinning around an imaginary line. This imaginary line, or **axis**, runs through the center of Earth from the North Pole to the South Pole. It takes 24 hours, or one day, for Earth to complete one **rotation** on its axis. This rotation causes day and night.

✔ **What are two motions of Earth?**

It takes one year for Earth to revolve once around the sun. At the same time, Earth rotates on its axis once each day.

March
February
January
December
November
October
September
August
July
June
May
April

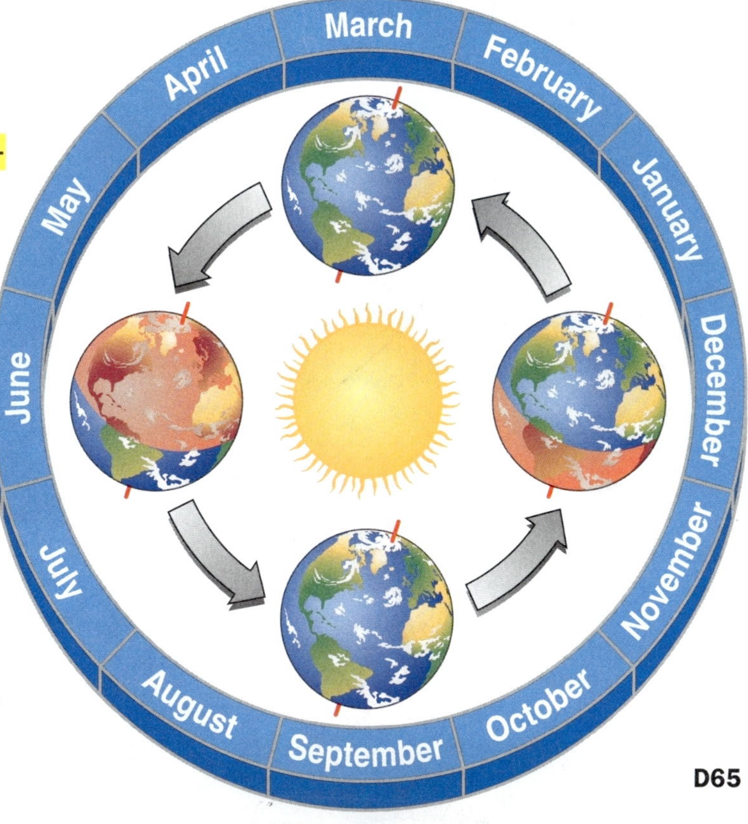

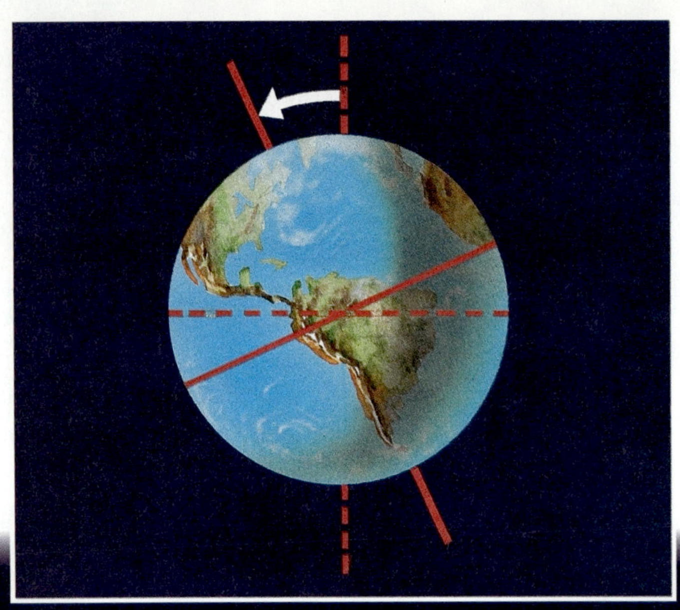

Earth and the Seasons

Earth's axis is slightly tilted. This tilt, along with the constant movement of Earth around the sun, causes seasons. For part of each year, the Northern Hemisphere tilts toward the sun. That part of Earth receives more direct energy from sunlight. We call this time of year summer. Six months later, when Earth has moved to the opposite side

SEPTEMBER 22 OR 23
Earth's axis is tilted neither toward nor away from the sun. All places on Earth have equal hours of daylight and night. Fall begins in the Northern Hemisphere, and spring begins in the Southern Hemisphere.

JUNE 21 OR 22
The Northern Hemisphere is pointed most directly toward the sun, causing more hours of daylight, shorter nights, and warmer weather. This marks the beginning of summer. The Southern Hemisphere has its shortest day and winter begins.

DECEMBER 21 OR 22
The Northern Hemisphere is tilted most directly away from the sun, causing fewer daylight hours, longer nights, and colder weather. This marks the beginning of winter. The Southern Hemisphere has its longest day and summer begins.

MARCH 20 OR 21
Earth's axis is tilted neither toward nor away from the sun. All places on Earth have equal hours of daylight and night. Spring begins in the Northern Hemisphere, and fall begins in the Southern Hemisphere.

of its orbit, the Northern Hemisphere tilts away from the sun. We call this time of year winter. That part of Earth receives less direct energy from sunlight.

✓ **When does winter occur in the Northern Hemisphere?**

Summary

The moon seems to change shape as it moves around Earth. These phases of the moon are caused by differences in the amount of lighted surface visible from Earth. As the moon orbits Earth, Earth is spinning on an imaginary axis. It takes one day for Earth to complete one rotation.

As Earth rotates it also revolves around the sun. It takes one year for Earth to complete a revolution. During its revolution some parts of Earth are titled toward the sun and other parts are tilted away from the sun. As a result, different locations on Earth receive different amounts of direct sunlight. This causes the seasons.

Review

1. Why would you classify the moon and Earth as satellites?
2. What are two ways that Earth moves?
3. What causes the seasons?
4. **Critical Thinking** How would day and night differ if Earth's axis were not tilted?
5. **Test Prep** What season would a town in the Southern Hemisphere have during January?
   A winter
   B spring
   C summer
   D fall

LINKS

MATH LINK

Solve a Problem The moon is about 384,000 kilometers from Earth. The average speed of an Apollo spacecraft was 38,600 kilometers per hour. At this rate, how long would it take to reach the moon?

WRITING LINK

Informative Writing—Identify the Phase Think about the characteristics of each phase of the moon. Write a letter to a friend, explaining how he or she can use observations of the moon to identify its phase.

SOCIAL STUDIES LINK

Apollo 11 On July 20, 1969, Neil Armstrong became the first human to set foot on the moon. Use reference materials to learn more about the historic flight of *Apollo 11*. Share your findings in a newspaper account of the mission.

TECHNOLOGY LINK

Visit the Harcourt Learning Site for related links, activities, and resources.
www.harcourtschool.com

WELCOME TO THE LEARNING SITE

LESSON 2

How Do Objects Move in the Solar System?

In this lesson, you can . . .

INVESTIGATE the ways planets move.

LEARN ABOUT our solar system.

LINK to math, writing, art, and technology.

INVESTIGATE

Planet Movement

Activity Purpose Even though you can't feel it, Earth moves through space at nearly 30 kilometers (about 19 mi) per second. At this speed, our planet moves around the sun almost 100 times as fast as most jet planes cruise. You can't make anything move that quickly. So, in this investigation you'll **make a model** that shows how the planets in our solar system move.

Materials

- index cards
- scissors
- black marker

Activity Procedure

1. Label one of the cards *Sun*. Label each of the other cards with the name of one of the planets shown in the data table on the next page.
2. Put all of the cards face down on a table and shuffle them. Have each person choose one card.
3. Use the data table to find out which planet is closest to the sun. Continue **analyzing the data** and **ordering** the cards until you have all the planets in the correct order from the sun.

◄ This is Stonehenge, an ancient rock structure located in Britain. Stonehenge may have been used to study and predict the movement of Earth around the sun.

Planets and Distances from the Sun			
Planet	**Average Distance from the Sun (millions of km)**	**Planet**	**Average Distance from the Sun (millions of km)**
Earth	150	Pluto	5900
Jupiter	778	Saturn	1429
Mars	228	Uranus	2871
Mercury	58	Venus	108
Neptune	4500		

4. In a gym or outside on a playground, line up in the order you determined in Step 3. (Picture A)

5. If you have a planet card, slowly turn around as you walk at a normal pace around the sun. Be sure to stay in your own path. Do not cross paths with other planets. After everyone has gone around the sun once, **record** your **observations** of the planets and their movements.

Picture A

Draw Conclusions

1. The sun and the planets that move around it are called the solar system. What is the order of the planets, starting with the one closest to the sun?
2. What did you **observe** about the motion of the planets?
3. **Scientists at Work** Why did you need to **make a model** to study how planets move around the sun?

Investigate Further Look again at the distances listed in the data table. How could you change your model to make it more accurate?

Process Skill Tip

Models are used when direct observations aren't possible. The huge size of the solar system is one reason scientists **make models** to study this complex collection of space objects.

Our Solar System

FIND OUT

- about the star we know as the sun
- the ways objects move in our solar system

VOCABULARY

solar system
star
planet
asteroid
comet

The Sun

In the investigation you made a model of our solar system. A **solar system** is a group of objects in space that move around a central star. Our sun is a **star**, a burning sphere (SFEER) of gases. This enormous fiery ball is more than 1 million kilometers (about 621,000 mi) in diameter. The sun is the largest object in our solar system. It is larger than the rest of the objects in the solar system put together.

The sun puts out a lot of energy in all directions. In fact, it is the source of almost all the energy in our solar system. Some of this energy reaches Earth as light, and some reaches it as heat.

Two features of the sun's surface are shown on this page. The dark areas, called *sunspots*, are cooler than the rest of the sun's surface and don't give off as much light. The red streams and loops of gases that shoot out from the sun are called *prominences* (PRAHM•ih•nuhn•suhs). These hot fountains often begin near a sunspot. They can be thousands of kilometers high and just as wide. Sunspots and prominences usually last for only a few days. Some can last for a few months.

✓ **What is the largest object in our solar system?**

The sun is the largest object in our solar system. The next largest object, Jupiter, is small compared to the sun. Earth is even smaller. ▼

The sun has 99.8 percent of the mass of our solar system. The planets, from nearest to farthest from the sun, are Mercury, Venus, Earth, Mars, Jupiter, Saturn, Uranus, Neptune, and Pluto. Each planet travels in its own orbit, a path around the sun. ►

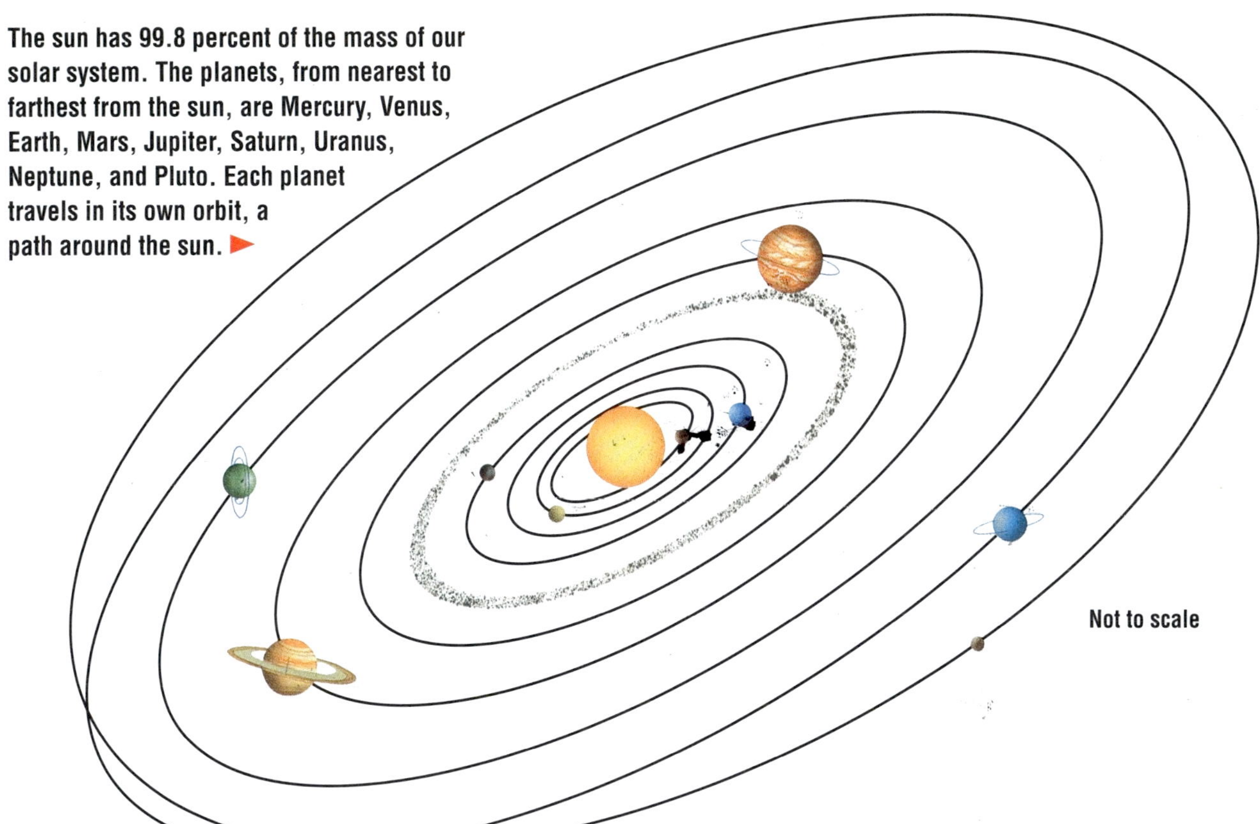

Other Objects in Our Solar System

As you saw in the investigation, our solar system is made up of the sun and nine planets. It also includes moons around the planets, and asteroids and comets.

A **planet**, such as Earth and its eight neighbors, is a large object that moves around a star. Most planets in our solar system also have at least one natural satellite, or moon. Earth and Pluto each have only one moon. Jupiter and Saturn, on the other hand, each have many moons.

Asteroids and comets are other objects that move around the sun. **Asteroids** are small and rocky. Most of them are scattered in a large area between the orbit paths of Mars and Jupiter. Some scientists hypothesize that these asteroids are pieces of a planet that never formed. All the asteroids put together would make an object less than half the size of Earth's moon.

A **comet** is a small mass of dust and ice that orbits the sun in a long, oval-shaped path. When a comet's orbit takes it close to the sun, some of the ice on the comet's surface changes to water vapor and streams out to form a long, glowing tail.

✓ **Name the objects in our solar system.**

◄ As a comet orbits the sun, its tail always points away from the sun. Comet Hale-Bopp passed near Earth in April 1997. Its orbit is so big that it will not be seen from Earth again for 2380 years.

Paths Around the Sun

In the investigation, student "planets" moved around a "sun." An object revolves as it moves around another object. The time for one complete orbit by a planet around the sun is its *year*. The orbits of the planets are not circles. Instead, they are a little bit elliptical, or oval, in shape.

Have you ever watched ice skaters spin? You may have noticed that as they bring their arms closer to their bodies, they spin faster. When they hold their arms out, they spin more slowly. The motion of the planets in their orbits is a little like the motion of the hands of a spinning ice skater. Planets with orbits closer to the sun move faster around the sun than those with orbits that are farther away.

✔ **Where in our solar system are the planets that orbit fastest?**

Earth's rotation on its axis causes day and night. The side of Earth that faces the sun has day. At the same time, the opposite side of Earth has night. ▲

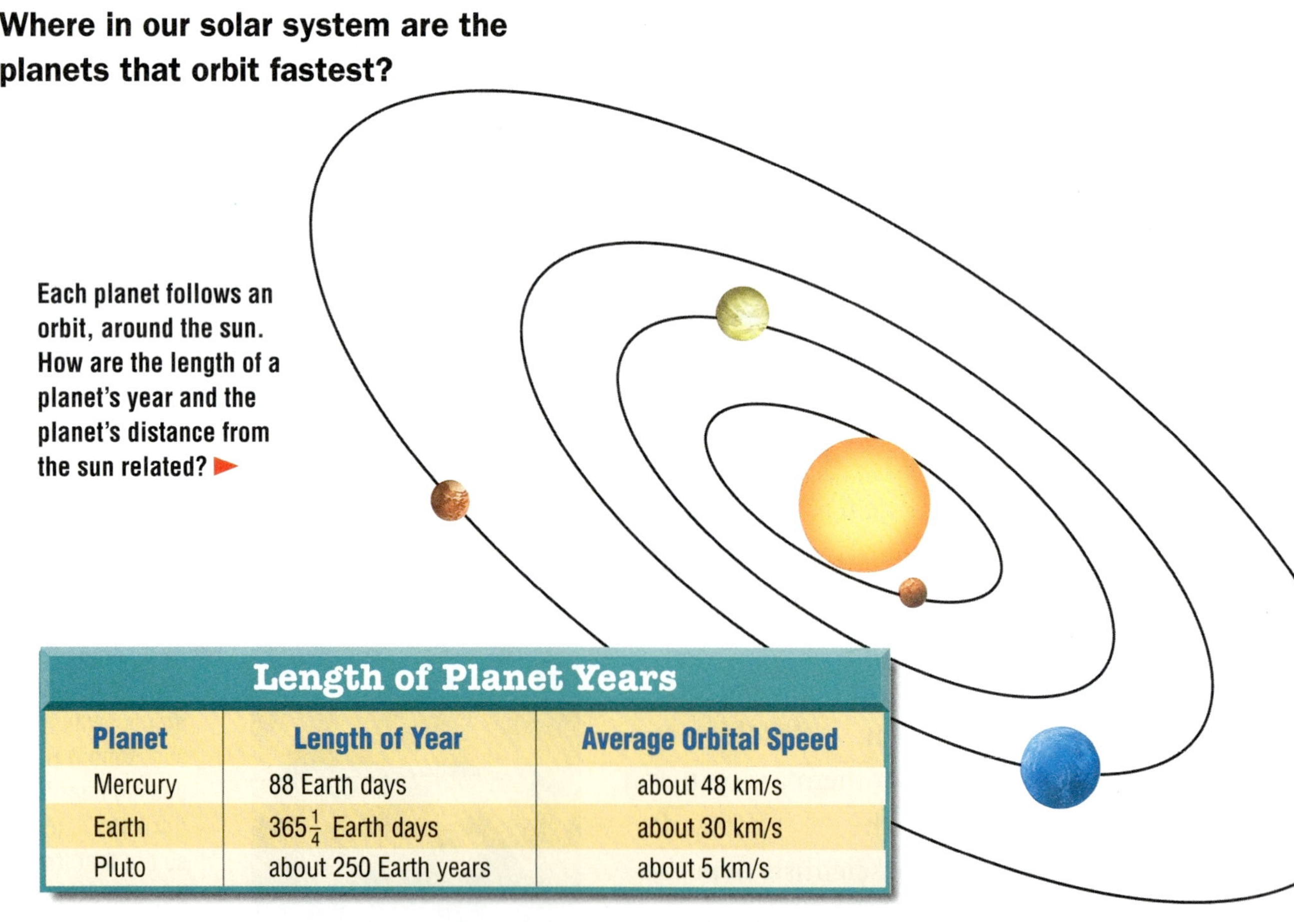

Each planet follows an orbit, around the sun. How are the length of a planet's year and the planet's distance from the sun related? ►

Length of Planet Years

Planet	Length of Year	Average Orbital Speed
Mercury	88 Earth days	about 48 km/s
Earth	$365\frac{1}{4}$ Earth days	about 30 km/s
Pluto	about 250 Earth years	about 5 km/s

Summary

Our solar system is made up of the sun, nine planets and their moons, asteroids, and comets. Each planet revolves in an elliptical orbit around the sun and rotates on its own axis.

Review

1. What is the sun?
2. In what way are the planets, comets, and asteroids alike?
3. How does a planet's distance from the sun affect its orbit speed?
4. **Critical Thinking** What might happen if the number of sunspots got much larger?
5. **Test Prep** What provides most of the heat and light to our solar system?
   A the comet Hale-Bopp
   B asteroids
   C the sun
   D Jupiter

LINKS

MATH LINK

Divide Whole Numbers The sun's diameter is about 1,392,000 km. About 109 Earths or 9 Jupiters would fit side by side across the sun. Use this information to figure out how many Earths would fit across Jupiter.

WRITING LINK

Expressive Writing—Poem Think about all the things you have learned about our solar system. Write a poem for your family describing how things move in our solar system or what makes up the solar system.

ART LINK

Drawing an Ellipse Tie the ends of a piece of string together to make a loop. Put the loop around two pushpins stuck several inches apart in a thick piece of cardboard. Place the point of a pencil inside the loop so the pencil touches the string. Keep the string tight as you draw around the pushpins. The shape you have drawn is an ellipse. What happens to the shape drawn if you move the pins closer together? Try it.

TECHNOLOGY LINK

Learn more about how asteroids move around the sun by visiting the National Air and Space Museum Internet site.
www.si.edu/harcourt/science

Smithsonian Institution®

LESSON 3

What Are the Planets Like?

In this lesson, you can . . .

INVESTIGATE distances between planets.

LEARN ABOUT the planets in our solar system.

LINK to math, writing, technology, and other areas.

Distances Between Planets

Activity Purpose If you've ever used a map, you know what a scale model is. A scale model is a way to compare large distances in a smaller space. In this investigation you will **use measurements** to **make a scale model** that shows the distances between planets in our solar system.

Materials

- piece of string about 4 m long
- meterstick
- 9 different-colored markers

Activity Procedure

1. Copy the chart shown on the next page.
2. At one end of the string, tie three or four knots at the same point to make one large knot. This large knot will stand for the sun in your model.
3. In the solar system, distances are often measured in astronomical units (AU). One AU equals the average distance from Earth to the sun. In your model, 1 AU will equal 10 cm. Use your meterstick to accurately measure 1 AU from the sun on your model. This point stands for Earth's distance from the sun. Use one of the markers to mark this point on the string. Note in your chart which color you used. (Picture A)

◀ Europa (yoo•ROH•puh) is one of Jupiter's many moons. This natural satellite has a diameter of 3100 kilometers (about 1925 mi) and takes about $3\frac{1}{2}$ Earth days to orbit Jupiter.

Planetary Distances from the Sun

Planet	Average Distance from the Sun (km)	Average Distance from the Sun (AU)	Scale Distance (cm)	Marker Color	Planet's Diameter (km)
Mercury	58 million	$\frac{4}{10}$	4		4876
Venus	108 million	$\frac{7}{10}$	7		12,104
Earth	150 million	1			12,756
Mars	228 million	2			6794
Jupiter	778 million	5			142,984
Saturn	1429 million	10			120,536
Uranus	2871 million	19			51,118
Neptune	4500 million	30			49,532
Pluto	5900 million	39			2274

4. Complete the Scale Distance column of the chart. Then measure and mark the position of each planet on the string. Use a different color for each planet, and **record** in your chart the colors you used.

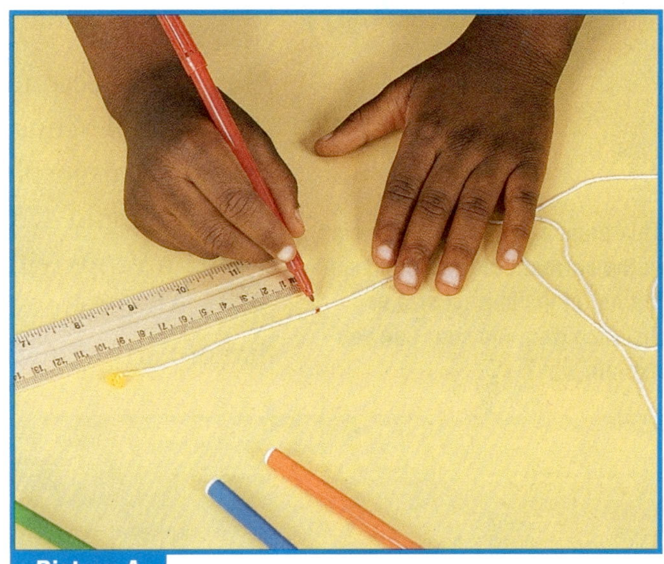
Picture A

Draw Conclusions

1. In your **model**, how far from the sun is Mercury? How far away is Pluto?
2. What advantages can you think of for using AUs to measure distances inside the solar system?
3. **Scientists at Work** Explain how it helped to **make a scale model** instead of trying to show actual distances between planets.

Investigate Further You can use a calculator to help make other scale models. The chart gives the actual diameters of the planets. Use this scale: Earth's diameter = 1 cm. Find the scale diameters of the other planets by dividing their actual diameters by Earth's diameter. Make a scale drawing showing the diameter of each planet.

Process Skill Tip

Models are made to study objects or events that are too small or too large to observe directly. A **scale model** shows large objects or areas in smaller sizes so that they can be more easily studied.

The Planets

FIND OUT

- about the planets in our solar system
- how moons and rings may have formed

VOCABULARY

inner planets
outer planets
gas giants

The Inner Planets

The area of the asteroid belt can be thought of as a dividing line between two groups of planets, the inner and outer planets. The **inner planets**—Mercury, Venus, Earth, and Mars—lie between the sun and the asteroid belt. Like the asteroids, the inner planets are rocky and dense. Unlike the asteroids, these planets are large and, except for Mercury, have atmospheres.

Mercury, the planet closest to the sun, is about the size of Earth's moon. Mercury, which is covered with craters, even looks like the moon. Very small amounts of some gases are present on Mercury, but there aren't enough of them to form an atmosphere.

Venus, the second planet from the sun, is about the same size as Earth. But Venus is very different from Earth. Venus is dry and has a thick atmosphere that traps heat. The temperature at the surface is about 475°C (887°F). The thick atmosphere presses down on Venus with a weight 100 times that of Earth's atmosphere. Also, Venus spins on its axis in a direction opposite from that of Earth's rotation.

This drawing shows the planets in the correct order from the sun but not at the correct size or distance from the sun. Can you explain why? ▼

Mercury has a diameter of 4876 kilometers (about 3031 mi) and is 58 million kilometers (about 36 million mi) from the sun. This inner planet has no moons and takes about 59 Earth days to make one rotation, or turn once on its axis. Mercury orbits the sun in about 88 Earth days.

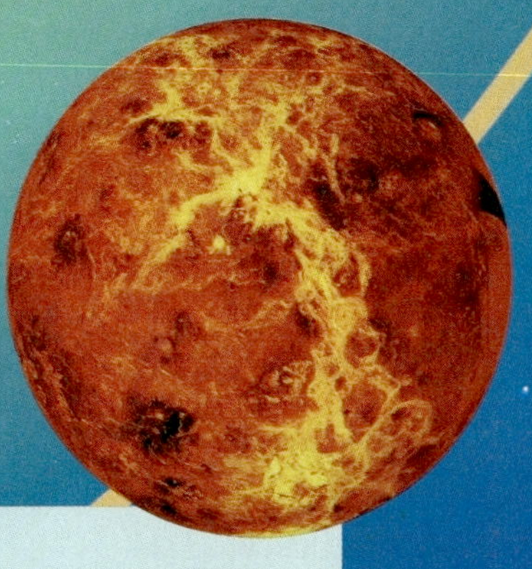

Venus has a diameter of 12,104 kilometers (about 7517 mi) and is 108 million kilometers (about 67 million mi) from the sun. This inner planet has no natural satellites. It takes Venus about 243 Earth days to make one rotation and 225 Earth days to orbit the sun.

Earth, the third planet from the sun, is the largest of the inner planets. It has one natural satellite, the moon. Earth is the only planet that has liquid water. It is also the only known planet that supports life. Earth's atmosphere absorbs and reflects the right amount of solar energy to keep the planet at the correct temperature for living things such as humans to survive.

Mars, the fourth planet from the sun, is sometimes called the Red Planet because its soil is a dark reddish brown. Mars has two moons and the largest volcano in the solar system—Olympus Mons (oh•LIHM•puhs MAHNS). Space probes have shown us that nothing lives on Mars. Dust storms can last for months and affect the whole planet. Although no liquid water has been found on Mars, it is believed that liquid water once existed there. This is because probes and satellites have found deep valleys and sedimentary rocks. These features probably were formed by flowing water.

✓ **List the inner planets in order from the sun.**

Earth has a diameter of 12,756 kilometers (about 7922 mi). Our planet is 150 million kilometers (about 93 million mi) from the sun. Earth has one moon and takes almost 24 hours to complete one rotation on its axis. It takes about 365 days to orbit the sun.

Not to scale

Mars has a diameter of 6794 kilometers (about 4230 mi) and is 228 million kilometers (about 142 million mi) from the sun. This inner planet has two moons. Mars completes a rotation in about 24.5 hours. Mars takes about 687 Earth days to complete one orbit around the sun.

The Outer Planets

On the other side of the asteroid belt are the **outer planets**—Jupiter, Saturn, Uranus, Neptune, and Pluto. Four of these planets—Jupiter, Saturn, Uranus, and Neptune—are large spheres made up mostly of gases. Because of this, these planets are often called the **gas giants**.

Jupiter is the largest planet in our solar system. A thin ring that is hard to see surrounds it. At least 16 moons orbit around it. Jupiter's atmosphere is very active. Its energy causes a circular storm known as the Great Red Spot. This weather, which is a lot like a hurricane, has lasted more than 300 years. It is so big around that three Earths would fit inside it.

Saturn is a gas giant known for its rings. Space probes have found that other planets also have rings. But Saturn's are so wide and so bright that they can be seen from Earth through a small telescope. Saturn has at least 18 named moons.

Uranus (YOOR•uh•nuhs), the seventh planet from the sun, is the most distant planet you can see without using a telescope. Uranus, a blue-green ball of gas and liquid, has at least 21 moons as well as faint rings around it.

Neptune, the farthest away of the gas giants, has at least eight moons and a faint ring. It also has circular storms, but none have lasted as long as the Great Red Spot on Jupiter.

In the investigation you saw that *Pluto* is the planet farthest from the sun. If you completed the Investigate Further, you also learned that Pluto is the smallest planet. From Pluto's surface the sun looks like a very bright star. Little heat or light reaches Pluto or its one moon. Unlike the other outer planets, Pluto is not a gas giant. Instead, Pluto has a rocky surface that is probably covered by frozen gases.

✓ **Which planets are gas giants?**

Uranus has a diameter of 51,118 kilometers (about 31,700 mi). This planet is 2870 million kilometers (1782 million mi) from the sun. Uranus makes one rotation in 17 hours and one orbit around the sun in about 84 Earth years.

Neptune has a diameter of 49,532 kilometers (about 30,740 mi) and is 4500 million kilometers (about 2795 million mi) from the sun. Neptune completes one rotation in a little more than 16 hours and one revolution in about 165 Earth years.

Pluto has a diameter of 2300 kilometers (about 1430 mi) and is 5900 million kilometers (about 3664 million mi) from the sun. It takes Pluto about 7 days to complete one rotation and 249 Earth years to complete one revolution.

▲ Titan, Saturn's largest moon, has a diameter of 5150 kilometers (about 3200 mi). Titan has no clouds in its atmosphere and is very cold.

▲ The rings around Saturn are made up of dust, ice crystals, and small bits of rock coated with frozen water. The rings are about 270,000 kilometers (about 167,700 mi) across but only about 10 kilometers (about 6 mi) thick.

Io (EYE•oh) is a moon of Jupiter. Io has a diameter of 3630 kilometers (about 2254 mi) and has several active volcanoes on its surface. The volcanoes are the big orange patches. ►

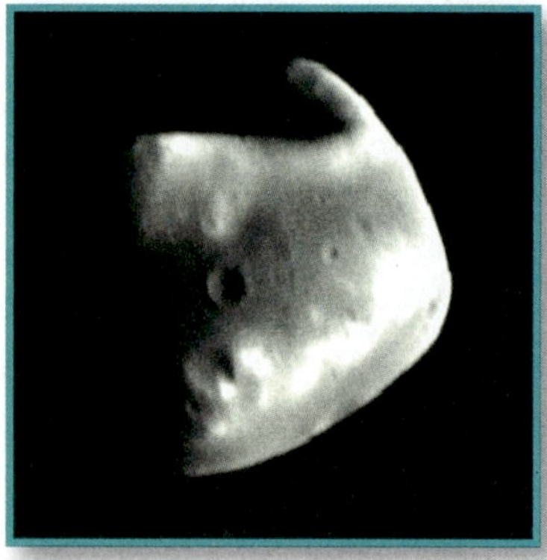

▲ Deimos (DY•muhs) is one of the two Martian moons. It has many craters and an uneven shape. It is about 14 kilometers (9 mi) in diameter.

Moons and Rings

Every planet except Mercury and Venus has at least one natural satellite, or *moon.* Earth's moon is round and rocky, and it has many craters. Others, like the two moons of Mars or the outer moons of Jupiter, are small and rocky and have uneven shapes. Jupiter and Mars orbit near the asteroid belt. So, their moons may be asteroids pulled in by the planets' gravity. The large moons of Jupiter and Saturn are almost like small planets. Io, one of Jupiter's larger moons, has active volcanoes. Titan (TYT•uhn), one of Saturn's moons, has a dense atmosphere that glows red-orange.

Besides having moons, each of the gas giants has a system of rings. These rings are made of tiny bits of dust, ice crystals, and small pieces of rock. Saturn's rings may have formed as a moon was pulled apart by gravity because it got too close to the planet.

✔ **What are planet rings made of?**

▲ Phobos (FOH•buhs) is the other moon that orbits Mars. Like Deimos, it is a small, rocky object. Its diameter is about 22 kilometers (14 mi). Phobos makes three trips around its planet each Martian day.

Summary

The inner planets—Mercury, Venus, Earth, and Mars—are small and rocky. Four of the five outer planets are gas giants. They are Jupiter, Saturn, Uranus, and Neptune. The outer planet that is farthest from the sun is Pluto, another rocky planet. Most of the planets have at least one moon. The gas giants also have rings.

Review

1. Name the inner planets, starting with the planet closest to the sun.
2. What can be thought of as the dividing line between the inner planets and the outer planets?
3. Which planets have no moons?
4. **Critical Thinking** Compare and contrast Venus and Earth.
5. **Test Prep** The outer planet that is **NOT** a gas giant is —
   A Jupiter
   B Saturn
   C Neptune
   D Pluto

LINKS

MATH LINK

Make a Bar Graph Make a bar graph showing the diameter of each planet. Use data from the table on page D75 and a computer program such as *Graph Links*.

WRITING LINK

Persuasive Writing—Opinion Suppose you are a real-estate agent trying to get adults to move to the planet of your choice. Write a newspaper ad pointing out all the benefits of living on your chosen planet.

ART LINK

View of a Planet Paint or draw a realistic landscape of the surface of another planet. Or paint the view of the planet as it would be seen from one of its moons.

LITERATURE LINK

The Wonderful Flight to the Mushroom Planet Read this book by Eleanor Cameron to find out what happens when two boys go on an adventure in space. Compare the planet with Earth.

TECHNOLOGY LINK

Learn more about planets and other objects in space by visiting this Internet site.

www.scilinks.org/harcourt

LESSON 4

How Do People Study the Solar System?

In this lesson, you can . . .

INVESTIGATE how to make a telescope.

LEARN ABOUT how people study objects in space.

LINK to math, writing, social studies, and technology.

◄ This vehicle is the moon rover. Astronauts used it to travel over the surface of Earth's only natural satellite, the moon.

Telescopes

Activity Purpose Have you ever looked up into the sky at night and wished you could see some of the objects more clearly? Because distances in space are so great, scientists need to use instruments to study what is beyond Earth's atmosphere. In this investigation you will make a simple telescope and use it to **observe** some objects in space.

Materials

- small piece of modeling clay
- 1 thin (eyepiece) lens
- small-diameter cardboard tube
- 1 thick (objective) lens
- large-diameter cardboard tube

CAUTION

Activity Procedure

1. Press small pieces of clay to the outside of the thin lens. Then put the lens in one end of the small tube. Use enough clay to hold the lens in place, keeping the lens as straight as possible. Be careful not to smear the middle of the lens with clay. (Picture A)
2. Repeat Step 1 using the thick lens and large tube.

Picture A

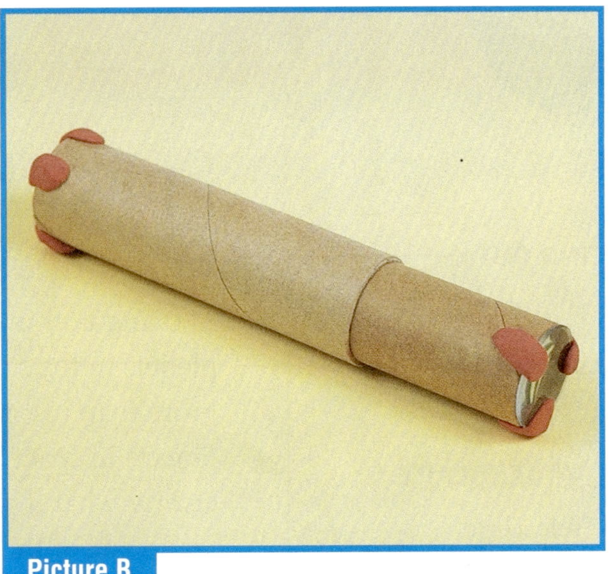
Picture B

3. Slide the open end of the small tube into the larger tube. You have just made a telescope. (Picture B)

4. Hold your telescope up, and look through one lens. Then turn the telescope around, and look through the other lens. CAUTION **Never look directly at the sun.** Slide the small tube in and out of the large tube until what you see is in focus, or not blurry. How do objects appear through each lens? **Record** your **observations.**

Draw Conclusions

1. What did you **observe** as you looked through each lens?
2. Using your observations, **infer** which lens you should look through to **observe** the stars. Explain your answer.
3. **Scientists at Work** Astronomers (uh•STRAWN•uh•merz) are scientists who study objects in space. Some astronomers use large telescopes with many parts to **observe** objects in space. How would your telescope make observing objects in the night sky easier?

Investigate Further Use your telescope to **observe** the moon at night. Make a list of the details you can see using your telescope that you can't see using only your eyes.

Process Skill Tip

When you **observe** an object, you use your senses to notice details about it. Using an instrument such as a telescope helps you observe objects that are too far away to be seen clearly using only your eyes.

Space Exploration

FIND OUT

- about telescopes
- about missions into space

VOCABULARY

telescope
space probe

Telescopes

Using nothing more than your eyes, you can see most of the planets in the solar system. But what if you want to see them as more than just points of light in the sky? What if you want to see objects in space that are even farther away than the visible planets? Or what if you want to see smaller objects such as moons? To do any of these things, you need to use a telescope. A **telescope** is a device people use to observe distant objects.

Two very different types of telescopes are used to observe objects in space: *radio telescopes* and *optical telescopes,* or telescopes that use light. There are also two types of optical telescopes. A refracting telescope, which is what you made in the investigation, uses lenses to magnify an object, or make it appear larger. A reflecting telescope uses a curved mirror to magnify an object. Most large telescopes used today are reflecting telescopes.

Even without using a telescope, you can see dark areas, bright areas, and some details on the moon. ▼

This telescope is very old. It uses two lenses to magnify objects that are far away. ▼

◀ Using a strong telescope lets you see more detail. Compare this photograph with the other one of the moon above.

▲ The large mirror of the Keck telescope is made up of many smaller mirrors. They work together to gather light and magnify images of objects in space.

▲ This is a photograph of Keck Observatory in Hawai`i. An *observatory* is a building where scientists study the planets, the sun, and other distant objects in the sky. This observatory is at the top of an inactive volcano more than 4000 meters (about 13,130 ft) above sea level.

Telescopes that scientists use are much larger than the one you made in the investigation. Besides being large, telescopes that are used to study space objects are powerful. Many of them have cameras that constantly take pictures of space. Computers keep the telescopes pointed at the same place in the sky. This allows the powerful lenses and mirrors to collect more light so they can produce brighter and clearer pictures. Astronomers then study the pictures to find out about space objects.

Earth's atmosphere limits what optical telescopes can "see." Moving air causes the "twinkling" of stars. It also blurs pictures taken using optical telescopes. This is why observatories that use optical telescopes are often located high on mountains.

Scientists have found that stars and other objects in space give off more than just light energy that we can see. Radio telescopes work the way optical telescopes do. But instead of collecting and focusing light, they collect and focus invisible radio waves. Moving air, clouds, and poor weather don't affect radio waves. Computers process the data collected by radio telescopes. The computers then make "pictures" that astronomers can study.

A radio telescope collects radio waves with a large, bowl-shaped antenna. Scientists study images formed by these waves to learn about the objects that gave them off. This radio telescope is in Arecibo (ar•uh•SEE•boh), Puerto Rico. ▶

The telescopes you have seen so far in this chapter are Earth-based, or located on Earth's surface. Scientists have also built telescopes for use in space. These telescopes don't have any problems caused by the atmosphere. The Inside Story shows the parts of the most famous space-based telescope, the Hubble Space Telescope.

✔ **What is a telescope?**

THE INSIDE STORY

HUBBLE SPACE TELESCOPE

The Hubble Space Telescope, or HST, is a reflecting telescope. Its mirror, which has a diameter of 240 cm (about 94 in.), can "see" details ten times as clearly as telescopes on Earth.

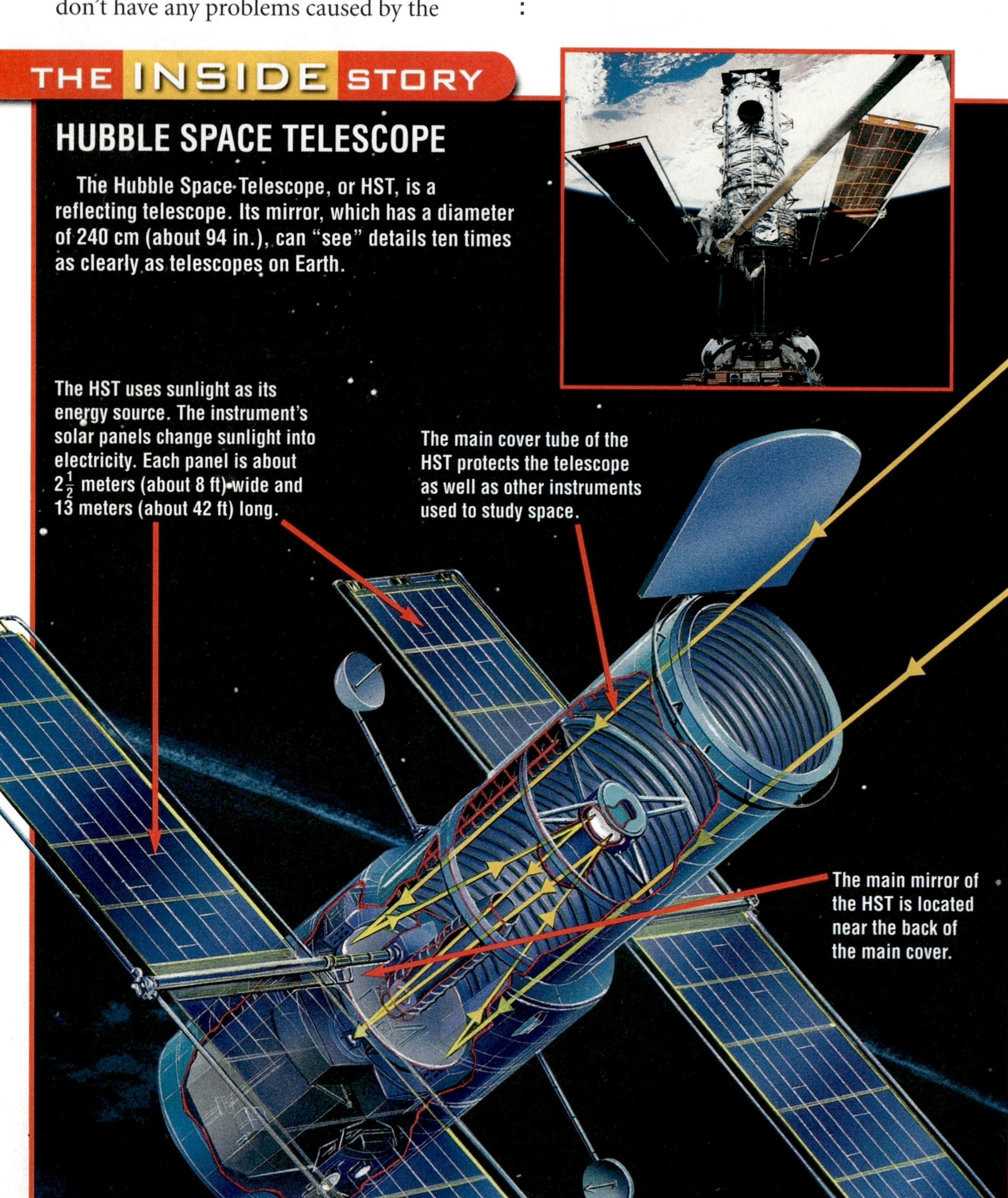

▲ During some of the Apollo missions, astronauts explored our moon's surface. This spacecraft allowed astronauts to land on the moon. Later they could blast off and return to the main part of their ship, which was circling the moon.

Mir was a Russian *space station* in orbit above Earth. Aboard *Mir*, astronauts and scientists from many countries worked for months at a time. They experimented to find out how conditions in space affect things and people. ▼

Crewed Missions

Another way to learn about space is to actually go there. Trips that people take into space are called *crewed missions.* Crewed missions are useful because people can actually find out for themselves what it is like to live and work in space.

The first person to be sent into space was a Russian, Yuri Gagarin, in 1961. Since then, astronauts from many countries have made trips into space. One of the most famous missions was carried out on *Apollo 11,* which was launched by the United States on July 16, 1969. On this mission Neil Armstrong and Edwin "Buzz" Aldrin spent two hours exploring the moon's surface. The United States sent five more crewed missions to the moon before the Apollo program ended in 1972.

Today the United States uses the space shuttle to carry crews, materials, and satellites to and from space. Astronauts on the shuttles do experiments, launch and get back satellites, and repair instruments.

✔ **What is a crewed space mission?**

Astronauts aboard the U.S. space shuttles also do experiments. These astronauts are working in Space Lab, which rides in the cargo area of a space shuttle. ▼

Space Probes

We have learned much of what we know about the solar system by using space probes. **Space probes** are vehicles that carry cameras, instruments, and other tools. Probes are sent to explore places that are too dangerous or too far away for people to visit. Probes gather data and send it back to Earth for study. The pictures of Mars and Callisto were sent to Earth by probes millions of kilometers away in space.

Some probes have fly-by missions. That is, they fly by the object to be studied but do not land. As they pass the object, they gather data, including pictures. Other probes land on the surfaces of planets. These probes take pictures, collect and analyze rock samples, test for the presence of substances such as water, and collect other data.

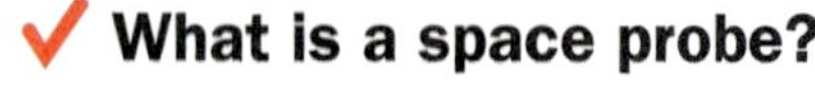

✔ **What is a space probe?**

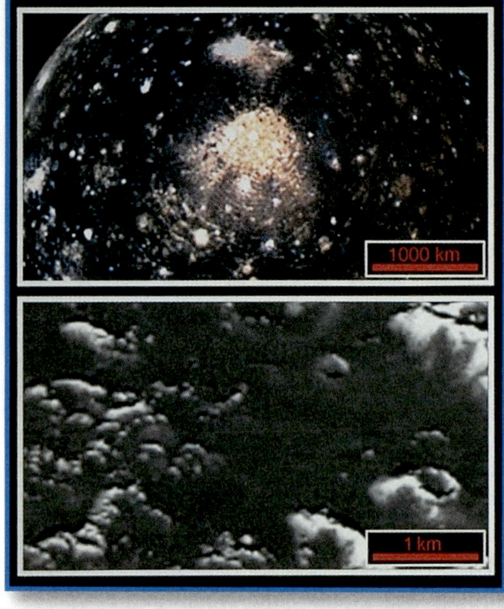

▲ Information about this moon of Jupiter, Callisto, was gathered as the space probe *Galileo* flew past it.

▲ The space probe *Galileo* was sent to Jupiter in 1989. Instruments launched from *Galileo* measured the sizes of particles that make up the clouds of Jupiter and the amounts of hydrogen and helium in Jupiter's atmosphere.

Sojourner is a probe that was used to study the surface of Mars. *Sojourner* took thousands of photos of the surface as it gathered information about the Red Planet. It landed as part of *Pathfinder* on July 4, 1997.▼

▲ *Voyager 2*, launched in 1977, flew by Jupiter in 1979, Saturn in 1981, and Uranus in 1986. The probe then headed on to Neptune.

Summary

A telescope is a device people use to observe distant objects. A refracting telescope uses lenses, and a reflecting telescope uses mirrors to magnify an object. Radio telescopes collect and focus radio waves. The Hubble Space Telescope is an optical telescope in orbit around Earth. Crewed missions and space probes are other ways to study objects in space.

Review

1. What are two types of telescopes?
2. What limits the things an Earth-based optical telescope can "see"?
3. What kind of work do astronauts on a crewed space mission do?
4. **Critical Thinking** What are the advantages of sending crewed missions instead of probes into space? The disadvantages?
5. **Test Prep** An instrument that uses lenses to magnify distant objects is a —
   A radio telescope
   B refracting telescope
   C reflecting telescope
   D space station

LINKS

MATH LINK

Solve a Problem When Earth and Jupiter are closest together, they are about 630 million km apart. *Voyager 2* took two years to reach Jupiter. About how far did it travel each year?

WRITING LINK

Informative Writing—Description Suppose you are an astronaut who will take part in the next space-shuttle mission. Find out about a typical day on the space shuttle. Then write a composition for a friend describing a typical day of your mission.

SOCIAL STUDIES LINK

International Space Station Find out which countries are building parts of the International Space Station. Locate each country on a map or globe. Make a chart that lists the name of each country, its continent, and which parts of the space station it is building.

TECHNOLOGY LINK

To learn more about pictures from space, watch *Hubble Images* on the **Harcourt Science Newsroom Video.**

Discovering the Planets

Five of the other eight planets in our solar system can be seen using just your eyes. These five—Mercury, Venus, Mars, Jupiter, and Saturn—were all known to astronomers before 1700. These early astronomers named the planets for gods and goddesses of Roman mythology.

Galileo and the Telescope

The invention of the telescope made astronomy a more complex science. Galileo, an Italian scientist and inventor who lived from 1564 to 1642, was the first person to use a telescope to view the stars and planets. Compared to telescopes now, his telescope wasn't very powerful. It magnified, or enlarged the view, only about 20 or 30 times. However, using just that weak telescope, Galileo was the first person to observe mountains on the moon, sunspots, and the phases of Venus. In 1610 Galileo was the first to see moons around Jupiter. His observations of planets and moons showed that not all bodies in space circled Earth, as many people at that time believed.

Discovering Uranus

Other scientists slowly improved on Galileo's telescope design. William Herschel, a British astronomer, discovered Uranus, the seventh planet from the sun. He used a telescope that his sister, Caroline, helped him build. On the night of March 13, 1781, Herschel saw something that he knew was not a star. He first thought it was a comet. Over time he mapped

Io, one of Jupiter's moons

The History of Planet Discoveries

Herschel and telescope

A.D. 1600 — A.D. 1700

Jupiter 1610
Jupiter's moons discovered

Uranus 1781
Uranus discovered

the orbit of the object and realized it was a planet, which was later named Uranus. For a long time, others had a hard time seeing what Herschel had seen, because their telescopes weren't as good.

Using Math to Find New Planets

Two people, the English astronomer John C. Adams and the French astronomer Urbain Leverrier (oor•BAN luh•vair•YAY), found the location of the eighth planet at about the same time. They did not see the planet themselves. Using mathematics, they predicted its location. Then they sent their predictions to other scientists.

Adams sent his prediction to the Astronomer Royal of England, who paid little attention to it. Leverrier sent his prediction to the Urania (oo•RAHN•ee•uh) Observatory in Berlin, Germany. The director there, Johann Galle (YOH•hahn GAH•luh), and his assistant used Leverrier's research and found Neptune on September 23, 1846.

Pluto was also discovered through mathematics. In 1905 American astronomer Percival Lowell noticed that something seemed to be affecting the orbits of Uranus and Neptune. He hypothesized that a ninth planet was the cause. He searched unsuccessfully for it until his death in 1916.

▲ These ink drawings are how Galileo recorded his telescope observations of Earth's moon.

In 1930 Pluto appeared as a small dot on three photographs Clyde Tombaugh took at Lowell Observatory, which was built by Percival Lowell.

THINK ABOUT IT

1. How have improvements in technology helped astronomers?
2. How did scientists Urbain Leverrier and Johann Galle work together?

Urania Observatory

A.D. 1800 — A.D. 1900 — A.D. 2000

Neptune 1846
Neptune discovered

Pluto 1930
Pluto discovered

Voyager 2 1989
Voyager 2 sends back photos of Neptune.

Pluto 1996
First detailed photos of Pluto

Today
Scientists use probes and orbiting telescopes to study faraway stars and to search for planets outside our solar system.

Lowell Observatory

Telescope photo of Pluto, 1930

Clyde Tombaugh

ASTRONOMER, INVENTOR

Tombaugh in 1995 with his third telescope

Working on a farm in Kansas taught Clyde Tombaugh to be persistent and creative with materials at hand. He was always interested in astronomy. His uncle and father had a telescope, which they gave him when he was 9 years old. By the time Tombaugh was 20, he decided to build his own telescope. He used part of a dairy machine for the base and part of his father's 1910 Buick.

Later, Tombaugh's uncle asked Tombaugh to build a telescope for him. Tombaugh also built a better telescope for him. With that homemade telescope, he observed Mars and Jupiter. He drew what he saw and sent his sketches to the Lowell Observatory in Flagstaff, Arizona. The scientists at the observatory were impressed by Tombaugh's drawings. They invited him to come to Arizona and work there. He stayed for 14 years.

On February 18, 1930, Tombaugh discovered the planet Pluto. Other scientists had predicted its existence, but he was the first to locate it in the sky. During his life, he discovered asteroids and hundreds of stars.

After completing college, Tombaugh taught navigation to U.S. Navy personnel during World War II. He also designed many new instruments, including an astronomy camera. He taught at New Mexico State University for nearly 20 years.

"I think the driving thing was curiosity about the universe. That fascinated me. I didn't think anything about being famous or anything like that, I was just interested in the concepts involved."

THINK ABOUT IT

1. What have you read about Tombaugh that leads you to think he was resourceful?
2. How was Tombaugh's discovery of Pluto related to work by other scientists?

ACTIVITIES FOR HOME OR SCHOOL

SUNDIAL

How can you make an instrument that uses the sun to tell time?

Materials

- small ball of clay
- short pencil
- cardboard, about 15 cm × 20 cm

Procedure

1. Use the lump of clay to stand the short pencil in the center of the cardboard. Make sure the sharpened end of the pencil is up.

2. Draw a half circle around the short pencil. The radius of the half circle should be the length of the pencil. The center of the half circle should be the lump of clay.
3. Put your sundial on a windowsill that gets sun all day. Each hour, trace the shadow of the pencil on the cardboard. Mark the time at the end of each line. Do this for six hours.
4. On the next sunny day, use your sundial to tell time.

Draw Conclusions

How does your sundial use Earth's movement to tell time?

SEASONS AND SUNLIGHT

How does the angle of light affect temperature?

Materials

- small 60-watt table lamp
- ruler
- graph paper
- black construction paper
- thermometer

Procedure

1. Darken the room. Place the lamp so that it shines directly down from 30 cm above a piece of graph paper. Draw an outline of the lighted area.
2. Repeat Step 1, this time placing the lamp so that it shines at an angle from the side.
3. Replace the graph paper with a sheet of black paper. Place the lamp as in Step 1. Then measure the temperature in the lighted area after 15 minutes.
4. Place the lamp as in Step 2. Measure the temperature in the lighted area after 15 minutes.

Draw Conclusions

How did moving the lamp change the area covered by the light? How did it affect the temperature? How is this activity related to the pictures shown on page D66?

CHAPTER 3 Review and Test Preparation

Vocabulary Review

Use the terms below to complete the sentences. The page numbers in ( ) tell you where to look in the chapter if you need help.

satellite (D64)	**planet** (D71)
phases (D64)	**asteroid** (D71)
orbit (D64)	**comet** (D71)
revolution (D65)	**inner planets** (D76)
rotation (D65)	**outer planets** (D78)
axis (D65)	**gas giants** (D78)
solar system (D70)	**telescope** (D84)
star (D70)	**space probe** (D88)

1. A ____ is a group of planets and their moons that orbit a central star.
2. Venus is a ____ that revolves around the sun.
3. An ____ is an imaginary line around which a planet rotates, or spins.
4. Any object orbiting another object is a ____.
5. The four planets nearest the sun are called the ____.
6. An ____ is a rocky object that orbits the sun in a path between Mars and Jupiter.
7. A ____ is a vehicle sent into space in order to explore places too dangerous or too far away for people to visit.
8. The path a planet takes around the sun is its ____.
9. A ____ is a burning ball of gases.
10. Neptune is one of the four ____.
11. A ____ is an instrument used to observe distant objects.
12. A ____ is a space object made of ice, dust, and gases.
13. The five planets on the outer side of the asteroid belt are called the ____.
14. The moon seems to have different ____, or shapes, during a cycle of about 28 days.
15. A year is the time it takes Earth to make one ____.
16. A day is the time it takes Earth to make one ____.

Connect Concepts

List the planets, and classify them as to their position in the solar system. Be sure to list them in the correct order from the sun.

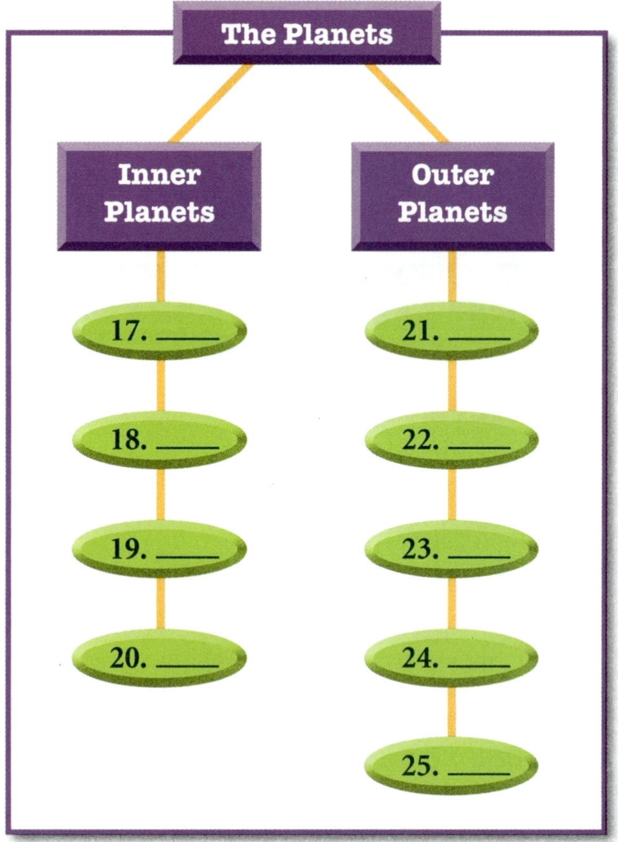

Check Understanding

Write the letter of the best choice.

26. The sun provides ____ of the energy to the solar system.

A most **C** none
B little **D** some

27. An imaginary line that runs through both poles of a planet is its —

F star **H** comet
G orbit **J** axis

28. ____ is the outer planet that is **NOT** a gas giant.

A Saturn **C** Uranus
B Pluto **D** Jupiter

29. This type of telescope never has problems seeing through Earth's atmosphere. It uses lenses to enlarge an object.

F Earth-based telescope
G reflecting telescope
H refracting telescope
J space-based telescope

30. Trips that people take into space are called —

A crewed missions
B uncrewed missions
C space probes
D observatories

Critical Thinking

31. Why do you think some stars in constellations look brighter than others?

32. Most asteroids are in the asteroid belt between the orbits of Mars and Jupiter. But some asteroids have "escaped" and have long, oval-shaped orbits like those of the comets. How do you think these asteroids escaped the asteroid belt?

Process Skills Review

33. In Lesson 1 you **observed** and compared the relative sizes of a basketball and a tennis ball. From a distance, you see buildings that seem to be the same height. What other **observations** would you make to confirm what you saw?

34. In Lesson 2 you **made a model** showing how planets rotate on their axes and revolve around the sun. How would you make a model to show the motions of Jupiter and its many moons?

35. In Lesson 3 you **made a model** to show the distances between the planets. What activity of planets would be modeled by cars racing around a racetrack?

36. In Lesson 4 you made a telescope to **observe** the night sky. Why did you need a telescope for this? What would you see if you didn't use a telescope?

Performance Assessment

Model Solar System

With a partner, draw a model of an imaginary solar system. Include one star, five planets, and two comets. Also include at least one more object that would be found in a solar system. Compare your model solar system with the one we live in.

UNIT D EXPEDITIONS

There are many places where you can learn about patterns on Earth and in space. You can learn about weather and the exploration of space by visiting the places below. You'll also have fun while you learn.

Fernbank Science Center

WHAT A science center that has a meteorology lab along with its other exhibits

WHERE Atlanta, Georgia

WHAT CAN YOU DO THERE? Tour the science center, see the exhibits, and visit the Meteorology Lab, where you can find current weather data.

U.S. Space & Rocket Center

WHAT A collection of rockets, spacecraft, and space exhibits

WHERE Huntsville, Alabama

WHAT CAN YOU DO THERE? Tour Rocket Park, visit the space exhibits, and explore the collection of hands-on exhibits and simulators.

GO ONLINE Plan Your Own Expeditions

If you can't visit the Fernbank Science Center or the U.S. Space & Rocket Center, visit a museum near you. Or log on to The Learning Site at **www.harcourtschool.com** to visit these science sites. Learn more about Earth and space.

References

Using Science Tools

Using a Hand Lens

A hand lens magnifies objects, or makes them look larger than they are.

1. Hold the hand lens about 12 centimeters (5 in.) from your eye.
2. Bring the object toward you until it comes into focus.

Using a Thermometer

A thermometer measures the temperature of air and most liquids.

1. Place the thermometer in the liquid. Don't touch the thermometer any more than you need to. Never stir the liquid with the thermometer. If you are measuring the temperature of the air, make sure that the thermometer is not in line with a direct light source.
2. Move so that your eyes are even with the liquid in the thermometer.
3. If you are measuring a material that is not being heated or cooled, wait about two minutes for the reading to become stable, or stay the same. Find the scale line that meets the top of the liquid in the thermometer, and read the temperature.
4. If the material you are measuring is being heated or cooled, you will not be able to wait before taking your measurements. Measure as quickly as you can.

Caring for and Using a Microscope

A microscope is another tool that magnifies objects. A microscope can increase the detail you see by increasing the number of times an object is magnified.

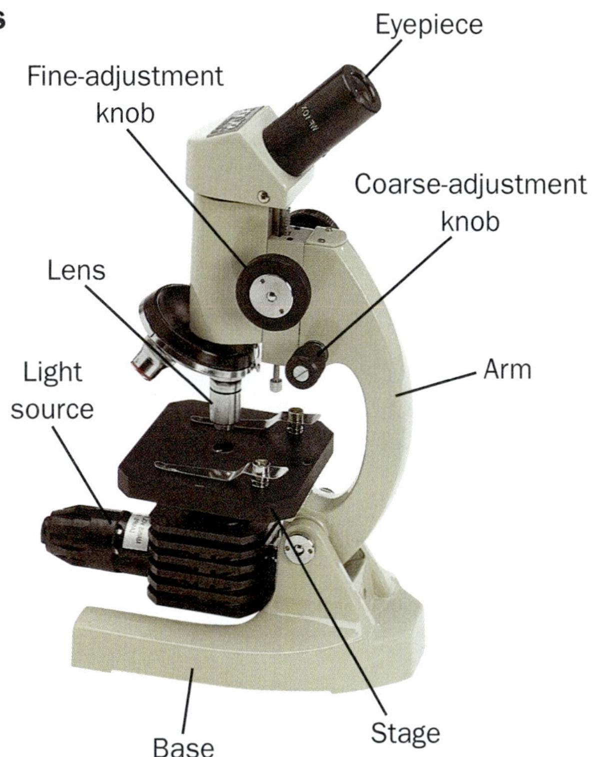

A Light Microscope

Caring for a Microscope

- Always use two hands when you carry a microscope.
- Never touch any of the lenses of a microscope with your fingers.

Using a Microscope

1. Raise the eyepiece as far as you can by using the coarse-adjustment knob. Place your slide on the stage.
2. Always start by using the lowest power. The lowest-power lens is usually the shortest. Start with the lens in the lowest position it can go without touching the slide.
3. Look through the eyepiece, and begin adjusting it upward with the coarse-adjustment knob. When the slide is close to being in focus, use the fine-adjustment knob.
4. When you want to use a higher-power lens, first focus the slide under low power. Then, watching carefully to make sure that the lens will not hit the slide, turn the higher-power lens into place. Use only the fine-adjustment knob when looking through the higher-power lens.

You may use a Brock microscope. This is a sturdy microscope that has only one lens.

1. Place the object to be viewed on the stage.
2. Look through the eyepiece, and begin raising the tube until the object comes into focus.

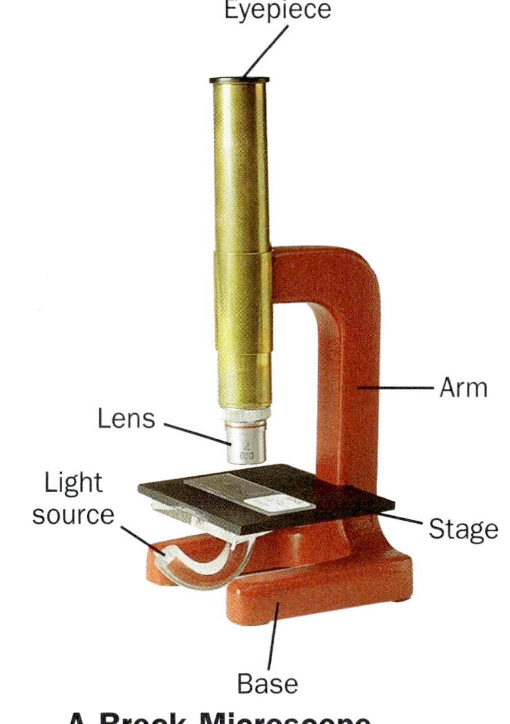

A Brock Microscope

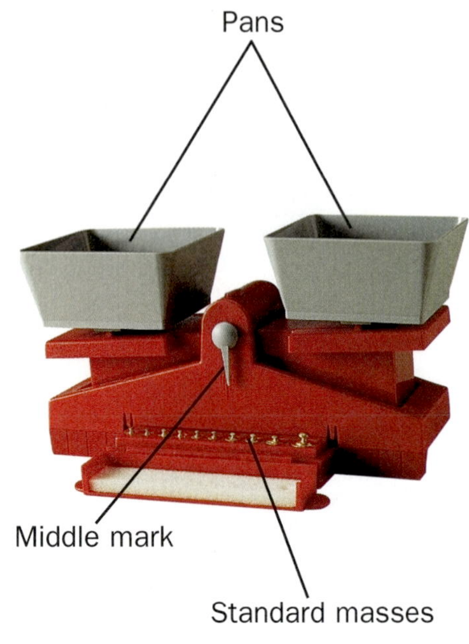

Using a Balance

Use a balance to measure an object's mass. Mass is the amount of matter an object has.

1. Look at the pointer on the base to make sure the empty pans are balanced.
2. Place the object you wish to measure in the left-hand pan.
3. Add the standard masses to the other pan. As you add masses, you should see the pointer move. When the pointer is at the middle mark, the pans are balanced.
4. Add the numbers on the masses you used. The total is the mass in grams of the object you measured.

Using a Spring Scale

Use a spring scale to measure forces such as the pull of gravity on objects. You measure weight and other forces in units called newtons (N).

Measuring the Weight of an Object

1. Hook the spring scale to the object.
2. Lift the scale and object with a smooth motion. Do not jerk them upward.
3. Wait until any motion of the spring comes to a stop. Then read the number of newtons from the scale.

Measuring the Force to Move an Object

1. With the object resting on a table, hook the spring scale to it.
2. Pull the object smoothly across the table. Do not jerk the object.
3. As you pull, read the number of newtons you are using to pull the object.

Beaker Graduate

Measuring Liquids

Use a beaker, a measuring cup, or a graduate to measure liquids accurately.

1. Pour the liquid you want to measure into a measuring container. Put your measuring container on a flat surface, with the measuring scale facing you.
2. Look at the liquid through the container. Move so that your eyes are even with the surface of the liquid in the container.
3. To read the volume of the liquid, find the scale line that is even with the surface of the liquid.
4. If the surface of the liquid is not exactly even with a line, estimate the volume of the liquid. Decide which line the liquid is closer to, and use that number.

Using a Ruler or Meterstick

Use a ruler or meterstick to measure distances and to find lengths of objects.

1. Place the zero mark or end of the ruler or meterstick next to one end of the distance or object you want to measure.
2. On the ruler or meterstick, find the place next to the other end of the distance or object.
3. Look at the scale on the ruler or meterstick. This will show the distance you want or the length of the object.

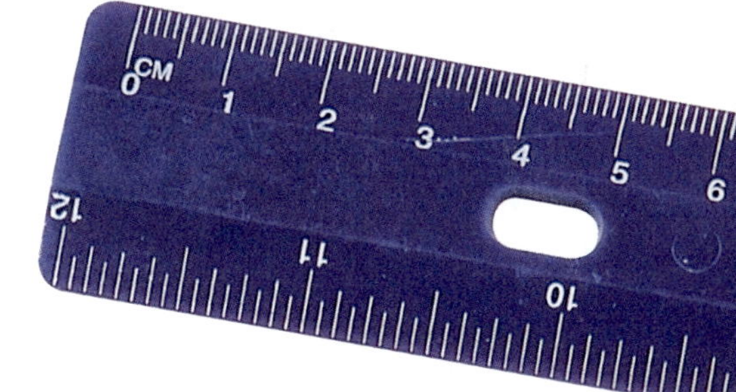

Using a Timing Device

Use a timing device such as a stopwatch to measure time.

1. Reset the stopwatch to zero.
2. When you are ready to begin timing, press *Start.*
3. As soon as you are ready to stop timing, press *Stop.*
4. The numbers on the dial or display show how many minutes, seconds, and parts of seconds have passed.

Visit the Multimedia Science Glossary to see illustrations of these words and to hear them pronounced.
www.harcourtschool.com/science

Glossary

As you read your science book, you will see words that may be new to you. The words have phonetic respellings to help you quickly know how to say them. In this Glossary you will see a different kind of respelling. Here, diacritical marks are used, as they are used in dictionaries. *Diacritical respellings* can show more exactly how words should sound.

When you see the ′ mark after a syllable, say that syllable more strongly than the other syllables. The page number after the meaning tells where to find the word in your book. The boldfaced letters in the Pronunciation Key show how each respelling symbol sounds.

PRONUNCIATION KEY

a	**a**dd, m**a**p	m	**m**ove, see**m**	u	**u**p, d**o**ne
ā	**a**ce, r**a**te	n	**n**ice, ti**n**	û(r)	b**ur**n, t**er**m
â(r)	c**are**, **air**	ng	ri**ng**, so**ng**	yo͞o	f**u**se, f**ew**
ä	p**a**lm, f**a**ther	o	**o**dd, h**o**t	v	**v**ain, e**v**e
b	**b**at, ru**b**	ō	**o**pen, s**o**	w	**w**in, a**w**ay
ch	**ch**eck, cat**ch**	ô	**o**rder, j**aw**	y	**y**et, **y**earn
d	**d**og, ro**d**	oi	**oi**l, b**oy**	z	**z**est, mu**s**e
e	**e**nd, p**e**t	ou	p**ou**t, n**ow**	zh	vi**s**ion, plea**s**ure
ē	**e**qual, tr**ee**	o͝o	t**oo**k, f**u**ll	ə	the schwa, an
f	**f**it, hal**f**	o͞o	p**oo**l, f**oo**d		unstressed vowel
g	**g**o, lo**g**	p	**p**it, sto**p**		representing the sound
h	**h**ope, **h**ate	r	**r**un, poo**r**		spelled
i	**i**t, g**i**ve	s	**s**ee, pa**ss**		*a* in **a***bove*
ī	**i**ce, wr**i**te	sh	**s**ure, ru**sh**		*e* in *sick***e***n*
j	**j**oy, le**dg**e	t	**t**alk, si**t**		*i* in *poss***i***ble*
k	**c**ool, ta**k**e	th	**th**in, bo**th**		*o* in *mel***o***n*
l	**l**ook, ru**l**e	t͟h	**th**is, ba**th**e		*u* in *circ***u***s*

Other symbols:

- • separates words into syllables
- ′ indicates heavier stress on a syllable
- ′ indicates light stress on a syllable

absorption [ab•sôrp′shən] The stopping of light when it hits a wall or other opaque object **(E106)**

abyssal plains [ə•bis′əl plānz′] Huge flat areas of ocean floor that are covered with thick layers of sediment **(D50)**

acceleration [ak•sel′ər•ā′shən] A change in the speed or direction of an object's motion **(F48)**

adaptation [ad′əp•tā′shən] A body part or behavior that helps an organism meet its needs in its environment **(A48)**

air mass [âr′mas′] A huge body of air which all has similar temperature and moisture **(D13)**

air pressure [âr′presh′ər] Particles of air pressing down on the Earth's surface **(D7)**

amplitude [am′plə•to͞od′] A measure of the strength of a sound wave; shown by height on a wave diagram **(E72)**

anthracite [an′thrə•sīt′] A hard, black rock; fourth stage of coal formation **(C55)**

artery [är′tər•ē] A blood vessel that carries blood away from the heart **(A105)**

arthropod [är′thrə•pod] An invertebrate with legs that have several joints **(A16)**

asteroid [as′tə•roid] A small rocky object that moves around the sun **(D71)**

atmosphere [at′məs•fir] The layer of air that surrounds our planet **(D6)**

axis [ak′sis] An imaginary line which runs through both poles of a planet **(D65)**

barometer [bə•rom′ət•ər] An instrument that measures air pressure **(D20)**

bituminous coal [bī•to͞o′mə•nəs kōl′] A fairly hard, dark brown or black rock; third stage of coal formation **(C55)**

brain [brān] The control center of your nervous system **(A110)**

buoyancy [boi′ən•sē] The ability of matter to float in a liquid or gas **(E20)**

camouflage [kam′ə•fläzh′] An animal's color or pattern that helps it blend in with its surroundings **(A52)**

capillary [kap′ə•ler′ē] A tiny blood vessel that allows gases and nutrients to pass from blood to cells **(A104)**

carbon dioxide [kär′bən dī•ok′sīd′] A gas breathed out by animals **(A72)**

cardiac muscle [kär′dē•ak mus′əl] A type of muscle that works the heart **(A99)**

cast [kast] A fossil formed when sediments or minerals fill a mold; it takes on the same outside shape as the living thing that shaped the mold **(C38)**

cell [sel] The basic building block of life **(A6)**

cell membrane [sel′ mem′brān] The thin layer that encloses and gives shape to a cell **(A7)**

cell wall [sel′ wôl′] A structure that keeps a cell rigid and provides support to an entire plant **(A8)**

charge [chärj] A measure of the extra positive or negative particles that an object has **(F6)**

chemical change [kem′i•kəl chānj′] A change that produces one or more new substances and may release energy **(E28)**

chemical reaction [kem′i•kəl rē•ak′shən] Another term for chemical change **(E28)**

chloroplast [klôr′ə•plast′] A part of a plant cell that contains chlorophyll, the green pigment plants need to make their food **(A8)**

circuit [sûr′kit] A path that is made for an electric current **(F12)**

cirrus [sir′əs] Wispy, high-altitude clouds that are made up of ice crystals **(D15)**

climate [klī′mit] The average temperature and rainfall of an area over many years **(A41, B28)**

comet [kom′it] A small mass of dust and ice that orbits the sun in a long, oval-shaped path **(D71)**

community [kə•myo͞o′nə•tē] All the populations that live in the same area **(B14)**

compression [kəm•presh′ən] The part of a sound wave in which air is pushed together **(E71)**

condensation [kon′dən•sā′shən] The process by which water vapor changes from a gas to liquid **(D34)**

conduction [kən•duk′shən] The transfer of thermal energy caused by particles of matter bumping into each other **(E49)**

conductor [kən•duk′tər] A material that electric current can pass through easily **(F13)**

conservation [kon′sər•vā′shən] The careful management and wise use of natural resources **(B68)**

consumer [kən•so͞o′mər] A living thing that eats other living things for energy **(B21)**

continental shelf [kon′tə•nen′təl shelf′] The ocean floor of the shore zone **(D49)**

convection [kən•vek′shən] The transfer of thermal energy by particles of a liquid or gas moving from one place to another **(E50)**

core [kôr] The dense center of Earth; a ball made mostly of two metals, iron and nickel **(C6)**

crater [krā′tər] A large basin formed at the top of a volcano when the top falls in on itself **(C22)**

crust [krust] Earth's outer layer; includes the rock of the ocean floor and large areas of land **(C6)**

cumulonimbus [kyo͞o′myo͞o•lō•nim′bəs] Towering, dark rain clouds with a nimbus, or halo, of gray-white **(D15)**

cumulus [kyo͞om′yə•ləs] Puffy cotton-ball clouds that begin to form when water droplets condense at middle altitudes **(D15)**

cytoplasm [sīt′ō•plaz′əm] A jellylike substance that fills most of the space in a cell **(A7)**

decomposer [dē′kəm•pōz′ər] A living thing that feeds on the wastes of plants and animals or on their remains after they die **(B21)**

deep ocean current [dēp′ ō′shən kûr′ənt] An

ocean current formed when cold water flows underneath warm water **(D44)**

density [den′sə•tē] The property of matter that compares the amount of matter to the space it takes up **(E14)**

dissolve [di•zolv′] To form a solution with another material **(E19)**

diversity [di•vûr′sə•tē] Variety **(B29)**

dormancy [dôr′mən•sē] State of much lower activity that some plants enter to survive colder weather **(A78)**

earthquake [ûrth′kwāk′] A vibration, or shaking, of Earth's crust **(C14)**

echo [ek′ō] A sound reflection **(E86)**

ecosystem [ek′ō•sis′təm] Groups of living things and the environment they live in **(B12)**

efficiency [i•fish′ən•sē] How well a machine changes effort into useful work **(F85)**

effort force [ef′ərt fôrs′] The force put on one part of a simple machine, for example, when you push or pull on a lever **(F70)**

electric cell [i•lek′trik sel′] A device that supplies energy to move charges through a circuit **(F12)**

electric current [i•lek′trik kûr′ənt] A flow of electric charges **(F12)**

electric field [i•lek′trik fēld′] The space around an object in which electric forces occur **(F8)**

electromagnet [i•lek′trō•mag′nit] An arrangement of wire wrapped around a core, producing a temporary magnet **(F25)**

embryo [em′brē•ō] A young plant **(A20)**

energy [en′ər•jē] The ability to cause a change **(E42)**

energy pyramid [en′ər•jē pir′ə•mid] A diagram that shows how much food energy is passed from one organism to another along a food chain **(B22)**

environment [in•vī′rən•mənt] Everything that surrounds and affects an animal, including living and nonliving things **(A40)**

epicenter [ep′i•sent′ər] The point on the surface of Earth that is right above the focus of an earthquake **(C15)**

esophagus [i•sof′ə•gəs] The tube that connects your mouth with your stomach **(A112)**

evaporation [ē•vap′ə•rā′shən] The process in which a liquid changes to a gas **(D34)**

fault [fôlt] A break in Earth's crust along which rocks move **(C14)**

fibrous roots [fī′brəs ro͞ots′] Long roots that grow near the surface **(A79)**

flowers [flou′ərz] Reproductive structures in flowering plants **(A22)**

focus [fō′kəs] The point underground where the movement of an earthquake first takes place **(C15)**

food web [fo͞od′ web′] A diagram that shows how food chains connect and overlap **(B23)**

force [fôrs] A push or pull **(F46)**

fossil [fos′əl] A preserved clue to life on Earth long ago **(C36)**

fossil fuel [fos′əl fyōō′əl] Fuel formed from the remains of organisms that lived long ago **(C52)**

frame of reference [frām′ uv ref′ər•əns] The things around you that you can sense and use to describe motion **(F41)**

friction [frik′shən] A force that keeps objects that are touching each other from sliding past each other easily **(F58)**

front [frunt] The border where two air masses meet **(D14)**

fruit [frōōt] The part of a flowering plant that surrounds and protects the seeds **(A22)**

fuel [fyōō′əl] A material that can burn **(E56)**

fulcrum [fŏŏl′krəm] The fixed point, or point that doesn't move, on a lever **(F70)**

fungi [fun′jī′] Living things that look like plants but cannot make their own food; for example, mushrooms **(A26)**

gas [gas] The state of matter that has no definite shape and takes up no definite amount of space **(E8)**

gas giants [gas′ jī′ənts] Planets which are large spheres made up mostly of gases—for example, Jupiter, Saturn, Uranus, and Neptune **(D78)**

germinate [jûr′mə•nāt′] To sprout; said of a seed **(A84)**

gravity [grav′ə•tē] A force that pulls all objects toward each other **(F56)**

greenhouse effect [grēn′hous′ i•fekt′] The warming of Earth caused by the atmosphere trapping thermal energy from the sun **(D12)**

habitat [hab′ə•tat′] An environment that meets the needs of an organism **(B20)**

heart [härt] The muscle that pumps blood through blood vessels to all parts of the body **(A105)**

heat [hēt] The transfer of thermal energy from one piece of matter to another **(E48)**

hibernation [hī′bər•nā′shən] A period when an animal goes into a long, deep "sleep" **(A59)**

humidity [hyōō•mid′ə•tē] The amount of water vapor in the air **(D21)**

hygrometer [hī•grom′ə•tər] A tool to measure moisture in the air **(D21)**

hyphae [hī′fē] Densely packed threadlike parts of a fungus **(A27)**

inclined plane [in′klīnd plān′] A flat surface with one end higher than the other **(F84)**

infrared radiation [in′frə•red′ rā′dē•ā′shən] The bundles of light energy that transfer heat **(E52)**

inner planets [in′ər plan′its] The planets closest to the sun; Mercury, Venus, Earth, and Mars **(D76)**

instinct [in′stingkt] A behavior that an animal begins life with **(A56)**

insulator [in′sə•lāt′ər] A material that current cannot pass through easily **(F13)**

intertidal zone [in′tər•tīd′əl zōn′] A narrow strip, along the shore, that is covered with water during high tide and exposed during low tide **(B36)**

invertebrate [in•vûr′tə•brit] An animal without a backbone **(A16)**

kinetic energy [ki•net′ik en′ər•jē] Energy of motion **(E42)**

large intestine [lärj′ in•tes′tən] The last part of the digestive system where water is removed from food **(A113)**

lava [lä′və] Melted rock that reaches Earth's surface **(C20)**

lever [lev′ər] A simple machine made up of a bar that turns on a fixed point **(F70)**

lignite [lig′nīt] A soft, brown rock; the second stage of coal formation **(C55)**

liquid [lik′wid] The state of matter that takes the shape of its container and takes up a definite amount of space **(E7)**

loudness [loud′nes] Your perception of the amount of sound energy reaching your ear **(E78)**

lungs [lungz] The main organs of the respiratory system **(A104)**

magma [mag′mə] Melted rock inside Earth **(C20)**

magma chamber [mag′mə chām′bər] An underground pool that holds magma, below a volcano **(C21)**

magnet [mag′nit] An object that attracts certain materials, such as iron or steel **(F18)**

magnetic field [mag•net′ik fēld′] The space all around a magnet where the force of the magnet can act **(F19)**

magnetic pole [mag•net′ik pōl′] The end of a magnet **(F18)**

mantle [man′təl] The thickest layer of Earth; found just below the crust **(C6)**

mass [mas] The amount of matter something contains **(E6)**

matter [mat′ər] Everything in the universe that has mass and takes up space **(E6)**

metamorphosis [met′ə•môr′fə•sis] The process of change; for example, from an egg to an adult butterfly **(A44)**

microorganisms [mī′krō•ôr′gən•iz′əmz] Organisms that are so small they can only be seen with a microscope; many have only one cell **(A9)**

mid-ocean ridge [mid′ō•shən rij′] A vast chain of mountains that runs along the centers of Earth's oceans **(D50)**

migration [mī•grā′shən] The movement of a group of one type of animal from one region to another and back again **(A57)**

mimicry [mim′ik•rē] An adaptation in which an animal looks very much like another animal or an object **(A52)**

mold [mōld] A common type of fungi that often look cottony or woolly **(A28)**

mold [mōld] A fossil imprint made by the outside of a dead plant or animal **(C38)**

motion [mō′shən] A change of position **(F40)**

natural gas [nach′ər•əl gas′] A gas, mostly methane, usually found with petroleum **(C53)**

near-shore zone [nir′shôr′ zōn′] Ocean zone that starts at the low-tide mark and goes out into the ocean **(B36)**

nerve [nûrv] A group of neurons that carries signals from the brain to the body and from the body to the brain **(A110)**

neuron [nŏŏr′on′] A nerve cell **(A110)**

newton [nōō′tən] The metric, or Système International (SI), unit of force **(F51)**

niche [nich] The role or part played by an organism in its habitat **(B21)**

nucleus [nōō′klē•əs] A cell's control center **(A7)**

nutrient [nōō′trē•ənt] A substance, such as a mineral, which all living things need in order to grow **(A72)**

opaque [ō•pāk′] Reflecting or absorbing all light; no image can be seen **(E106)**

open-ocean zones [ō′pən•ō′shən zōnz′] The deep parts of the oceans, located far from shore **(B36)**

orbit [ôr′bit] The path that an object such as a planet makes as it revolves around a second object **(D64)**

organ [ôr′gən] A group of tissues of different kinds working together to perform a task **(A98)**

outer planets [ou′tər plan′its] The planets farthest from the sun; Jupiter, Saturn, Uranus, Neptune, and Pluto **(D78)**

oxygen [ok′si•jən] One of the many gases in air **(A41)**

parallel circuit [par′ə•lel sûr′kit] A circuit that has more than one path along which current can travel **(F14)**

peat [pēt] A soft, brown material made up of partly decayed plants; first stage of coal formation **(C55)**

petroleum [pə•trō′lē•əm] A thick brown or black liquid fossil fuel; crude oil **(C53)**

phase [fāz] One of the different shapes the moon seems to have as it orbits around Earth **(D64)**

photosynthesis [fōt′ō•sin′thə•sis] The process by which a plant makes its own food **(A73)**

physical change [fiz′i•kəl chānj′] Any change in the size, shape, or state of a substance **(E26)**

pistil [pis′təl] A flower part that collects pollen **(A85)**

pitch [pich] A measure of how high or low a sound is **(E79)**

planet [plan′it] A large object that moves around a star **(D71)**

plate [plāt] Continent-sized slab of Earth's crust and upper mantle **(C8)**

pollination [pol′ə•na′shən] Transfer of pollen from a stamen to a pistil by wind or animals **(A85)**

population [pop′yōō•lā′shən] A group of the same species living in the same place at the same time **(B13)**

position [pə•zish′ən] A certain place **(F40)**

precipitation [prē•sip′ə•tā′shən] Water that falls to Earth as rain, snow, sleet, or hail **(D35)**

preservation [prez′ər•vā′shən] The protection of an area **(B72)**

prism [priz′əm] A solid object that bends light; not a lens **(E110)**

producer [prə•dōōs′ər] A living thing, such as a plant, that makes its own food **(B21)**

pulley [pŏŏl′ē] A simple machine made up of a rope or chain and a wheel around which the rope or chain fits **(F78)**

radiation [rā′dē•ā′shən] The bundles of energy that move through matter and through empty space **(E52)**

reclamation [rek′lə•mā′shən] The repairing of some of the damage done to an ecosystem **(B63)**

redesign [rē′di•zīn′] Changing the design of packaging or products in order to use fewer resources **(B71)**

reflection [ri•flek′shən] The bouncing of light off an object **(E102)**

refraction [ri•frak′shən] The bending of the path of light when it moves from one kind of matter to another **(E104)**

relative motion [rel′ə•tiv mō′shən] A motion that is described based on a frame of reference **(F41)**

resistor [ri•zis′tər] A material that resists the flow of current but doesn't stop it **(F13)**

revolution [rev′ə•lōō′shən] The movement of any object in an orbit, such as Earth moving around the sun **(D65)**

rotation [rō•tā′shən] The motion of a planet or other object as it turns on its axis **(D65)**

salinity [sə•lin′ə•tē] The amount of salt in water **(B30)**

satellite [sat′ə•līt′] An object that moves around another object in space; the moon is a satellite of Earth **(D64)**

screw [skrōō] An inclined plane wrapped around a pole **(F86)**

seismograph [sīz′mə•graf′] An instrument that records earthquake waves **(C16)**

series circuit [sir′ēz sûr′kit] A circuit that has only one path for current **(F14)**

shelter [shel′tər] A place where an animal is protected from other animals or from the weather **(A43)**

shore zone [shôr′ zōn′] The place where land and ocean meet **(D49)**

simple machine [sim′pəl mə•shēn′] One of the basic machines that make up other machines **(F70)**

small intestine [smôl′ in•tes′tən] A long tube of muscle where most food is digested **(A112)**

smooth muscle [smōōth′ mus′əl] A type of muscle found in the walls of some organs such as the stomach, intestines, blood vessels, and bladder **(A99)**

solar energy [sō′lər en′ər•jē] The energy given off by the sun **(E57)**

solar system [sō′lər sis′təm] A group of objects in space that move around a central star **(D70)**

solid [sol′id] The state of matter that has a definite shape and takes up a definite amount of space **(E6)**

solubility [sol′yōō•bil′ə•tē] A measure of the amount of a material that will dissolve in another material **(E19)**

solution [sə•lōō′shən] A mixture in which the particles of different kinds of matter are mixed evenly with each other and particles do not settle out **(E18)**

sonic boom [son′ik bōōm′] A shock wave of compressed sound waves produced by an object moving faster than sound **(E88)**

sound [sound] A series of vibrations that you can hear **(E70)**

sound wave [sound′ wāv′] A moving pattern of high and low pressure that you can hear **(E71)**

space probe [spās′ prōb′] An uncrewed space vehicle that carries cameras, instruments, and other research tools **(D88)**

speed [spēd] A measure of an object's change in position during a unit of time; for example, 10 meters per second **(F42)**

speed of sound [spēd′ uv sound′] The speed at which a sound wave travels through a given material **(E84)**

spinal cord [spī′nəl kôrd′] The tube of nerves that runs through your spine, or backbone **(A110)**

spore [spôr] A tiny cell that ferns and fungi use to reproduce **(A27, A85)**

stability [stə•bil′ə•tē] The condition that exists when the changes in a system over time cancel each other out **(B8)**

stamen [stā′mən] A flower part that makes pollen **(A85)**

star [stär] A huge, burning sphere of gases; for example, the sun **(D70)**

static electricity [stat′ik ē′lek•tris′i•tē] An electric charge that stays on an object **(F6)**

stomach [stum′ək] A bag made up of smooth muscles that mixes food with digestive juices **(A112)**

storm surge [stôrm′ sûrj′] A very large series of waves caused by high winds over a large area of ocean **(D41)**

stratosphere [strat′ə•sfir′] The layer of atmosphere that contains ozone and is located above the troposphere **(D8)**

stratus [strā′təs] Dark gray clouds that form a low layer and sometimes bring light rain or snow showers **(D15)**

striated muscle [strī′āt•ed mus′əl] A muscle with light and dark stripes; a muscle you can control by thinking **(A100)**

succession [sək•sesh′ən] The process that gradually changes an existing ecosystem into another ecosystem **(B52)**

surface current [sûr′fis kûr′ənt] An ocean current formed when steady winds blow over the surface of the ocean **(D44)**

system [sis′təm] A group of parts that work together as a unit **(B6)**

taproot [tap′rōōt′] A plant's single main root that goes deep into the soil **(A79)**

telescope [tel′ə•skōp′] A device people use to observe distant objects with their eyes **(D84)**

temperature [tem′pər•ə•chər] A measure of the average energy of motion of the particles in matter **(E43)**

thermal energy [thûr′məl en′ər•jē] The energy of the random motion of particles in matter **(E42)**

tide [tīd] The daily changes in the local water level of the ocean **(D42)**

tissue [tish′o͞o] A group of cells of the same type **(A98)**

trace fossil [trās′ fos′əl] A fossil that shows changes that long-dead animals made in their surroundings **(C37)**

translucent [trans•lo͞o′sənt] Allowing some light to pass through; blurry image can be seen **(E106)**

transparent [trans•pâr′ənt] Allows most light to pass through; clear image can be seen **(E106)**

transpiration [tran′spə•rā′shən] The giving off of water vapor by plants **(A78)**

trenches [trench′əz] Valleys that form on the ocean floor where two plates come together; the deepest places in the oceans **(D50)**

troposphere [trō′pə•sfir′] The layer of atmosphere closest to Earth **(D8)**

tuber [to͞o′bər] A swollen underground stem **(A87)**

vein [vān] A large blood vessel that returns blood to the heart **(A105)**

vent [vent] In a volcano, the rocky opening through which magma rises toward the surface **(C20)**

vertebrate [vûr′tə•brit] An animal with a backbone **(A16)**

visible spectrum [viz′ə•bəl spek′trəm] The range of light energy that people can see **(E110)**

volcano [vol•kā′nō] A mountain that forms when red-hot melted rock flows through a crack onto Earth's surface **(C20)**

volume [vol′yo͞om] The amount of space that matter takes up **(E13)**

water cycle [wôt′ər sī′kəl] The constant recycling of water on Earth **(D34)**

wave [wāv] An up-and-down movement of water **(D40)**

wavelength [wāv′length] The distance from one compression to the next in a sound wave **(E72)**

wedge [wej] A machine made up of two inclined planes placed back-to-back **(F88)**

weight [wāt] A measure of the force of gravity upon an object **(F57)**

wheel and axle [hwēl′ and ak′səl] A simple machine made up of a large wheel attached to a smaller wheel or rod **(F80)**

work [wûrk] That which is done on an object when a force moves the object through a distance **(F74)**

F

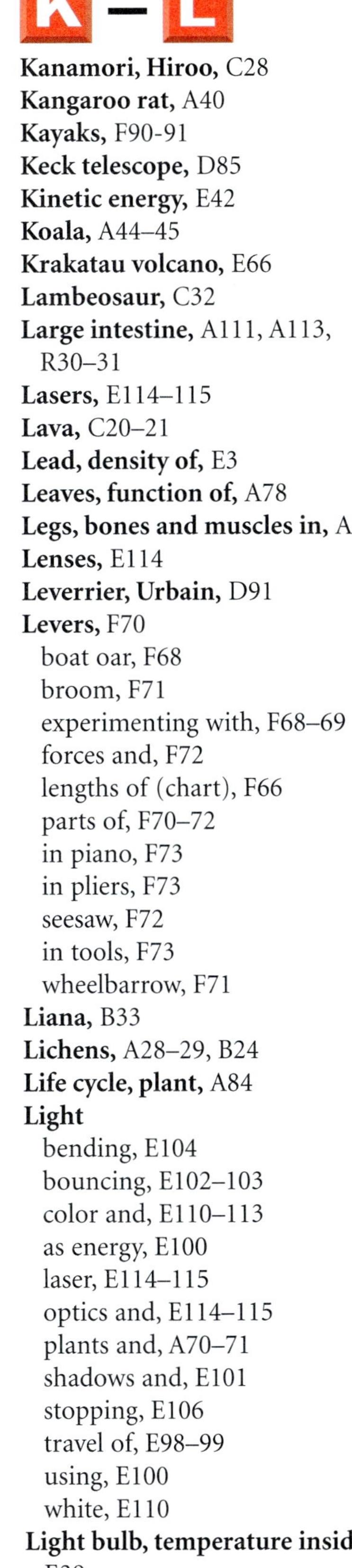

K–L

M

Q – R

S

PHOTO CREDITS:

Page Placement Key: (t)-top (c)-center (b)-bottom (l)-left (r)-right (fg)-foreground (bg)-background

Cover and Title Pages

Gary Neil Corbett/Superstock

Contents

Page: iv (fg) Steinhart Aquarium/Tom McHugh/Photo Researchers; iv (bg) Mark Lewis/Liaison International; v (fg) Steve Kaufman/DRK Photos; v (bg) H. Richard Johnson/FPG International; vi (fg) Francois/Gohier/Photo Researchers; vii (fg) Dorling Kindersley; vii (bg) FPG International; viii (fg); Dennis Yankus/Superstock; vii (bg) Pierre-Yves Goavec/Image Bank; ix (fg) Steve Berman/Liaison International; ix (bg) Photone Disk #50.

Unit A

Unit A Opener (fg) Steinhart Aquarium/Tom McHugh/Photo Researchers; (bg) Mark Lewis/Liaison International; A2 (i) Dr. Jeremy Burgess/Science Photo Library/Photo Researchers; A2-A3 (bg) Jeffrey L. Rotman Photography; A3 (ti) Tom Volk/University of Wisconsin-La Crosse; A4 M.I. Walker/Photo Researchers; A6 (r) Leonard Lee Rue III/Photo Researchers; A6 (i) Ray Simons/Photo Researchers; A6 (l), A6 (li) Dave G. Houser/Corbis; A7 Dr. Gopal Murti/Phototake; A8 (b) Biophoto Associates/Science Source/Photo Researchers; A9 (bg) Jeff Lepore/Photo Researchers; A9 (t) Michael Abbey/Photo Researchers; A9 (cr) Dr. E. R. Degginger/Color-Pic; A9 (cl) Biophoto Associates/Photo Researchers; A10 Dwight R. Kuhn; A12 Alex Kerstitch/Bruce Coleman, Inc.; A14 Graeme Teague; A15 (t) Ed Reschke/Peter Arnold, Inc.; A15 (b) J.C. Carton/Bruce Coleman, Inc.; A16 (t) William E. Ferguson; A16 (c) Dr. E. R. Degginger/Color-Pic; A16 (bl) Larry Miller/Photo Researchers; A16 (br) Dwight R. Kuhn; A17 David Dennis/Animals Animals; A18 Bumann/StockFood America; A20 William H. Allen, Jr.; A20 (i) Biophoto Associates/Photo Researchers; A21 (t), A21 (ci) Dwight R. Kuhn; A21 (ti) Richard V. Procopio/PictureQuest; A21 (b), A21 (bi) William E. Ferguson; A22 (t) Richard Shiell; A22 (ti) Bill Johnson; A22 (bi) J&L Waldman/Bruce Coleman, Inc.; A22 (bl), A22 (br) Dr. E.R. Degginger/Color-Pic, Inc.; A22 (ci) Bill Beatty/Wild & Natural; A23 (t) David R. Frazier; A23 (ti) Lance Beeny; A23 (bi) Dwight R. Kuhn; A24 Fritz Polking/Peter Arnold, Inc.; A26 Dr. Jeremy Burgess/Science Photo Library/Photo Researchers; A27 (l) Dr. E. R. Degginger/Color-Pic; A27 (r) W. Wayne Lockwood, M.D./Corbis; A28 (t) Chris Hellier/Corbis; A28 (c) Dr. E. R. Degginter/Color-Pic; A28 (bl) Noble Proctor/Photo Researchers; A28 (bi) Robert and Linda Mitchell/Mitchell Photography; A29 Phil Degginger/Color-Pic; A30 (l) Kevin Collins/Visuals Unlimited; A30 (r) Alfred Pasieka/Peter Arnold, Inc.; A30 (bg) Dr. Dennis Kunkel/Phototake; A31 Barbara Wright/Animals Animals; A32 (i) The Center for Environmental Biotechnology; A32 Andre Jenny/Focus Group/PictureQuest; A36-A37 Manoj Shah/Stone; A37 (t) Rudy Kuiter/Innerspace Visions; A37 (b) Peter Weimann/Animals Animals; A38 Corel; A40 (b) John Cancalosi/DRK; A40 (tl) Larry Minden/Minden Pictures; A40 (bg) Rich Reid/Earth Scenes; A40 (tr) Renee Lynn/Photo Researchers; A41 Daniel J. Cox/ Natural Exposures; A42 (t) T. Kitchen/Natural Selection; A42 (bl) Daniel J. Cox/Natural Selection; A42-A43 (b) Bios/Peter Arnold, Inc.; A43 (t) David E. Myers/Stone; A43 (bl) Osolinski, S. OSF/Animals Animals; A43 (li) Stephen Krasemann/Stone; A43 (ri) E & P Bauer/Bruce Coleman, Inc.; A44 (l) Ralph Clevenger/Westlight; A44 (bg) B & C Alexander/Photo Researchers; A44 (i) Dan Suzio Photography; A45 (t) Ben Simmons/The Stock Market; A45 (i) D. Parer & E. Parer-Cook/Auscape; A46 (b) Joe McDonald/Bruce Coleman, Inc.; A48 (l) Robert Lankinen/The Wildlife Collection; A48 (r) Martin Harvey/The Wildlife Collection; A48 (c) Zefa Germany/The Stock Market; A49 (t) Fritz Polking/Dembinsky Photo Associates; A49 (b) Tui De Roy/Minden Pictures; A49 (i) Zig Leszczynski/Animals Animals; A50 (l) Tom and Pat Leeson; A50 (r) Zig Leszczynski/Animals Animals; A50 (c) Fred Bavendam/Peter Arnold, Inc.; A50-A51 (b) Stuart Westmorland/Stone; A51 (t) Bruce Wilson/Stone; A51 (c) Wolfgang Kaehler Photography; A51 (bl) Martin Harvey/The Wildlife Collection; A51 (br) Bruce Davidson/Animals Animals; A52 (ti) Bruce Wilson/Stone; A52 (t) Art Wolfe/Stone; A52 (bl) Stephen Krasemann/Stone; A52 (br) Stouffer Prod./Animals Animals; A53 (l) Lior Rubin/Peter Arnold, Inc.; A53 (r) John Shaw/Bruce Coleman, Inc.; A54 C. Bradley Simmons/Bruce Coleman, Inc.; A56-A57 Mike Severns/Stone; A58 (t) Grant Heilman Photography; A58-A59 (b) Daniel J. Cox/Stone; A59 (t) Joe McDonald/Animals Animals; A59 (bl) Darrell Gulin/Stone; A59 (br) J. Foott/Bruce Coleman, Inc.; A60 (t) Tom Brakefield/The Stock Market; A60 (b) Mark Petersen/Stone; A61 (t) Darryl Torckler/Stone; A62 Katsumi Kasahara/Associated Press; A63 Phil McCarten/PhotoEdit; A64 Michael K. Nichols/NGS Image Collection; A67 (l) Ralph Clevenger/Westlight; A67 (li) Dan Suzio Photography; A68-69 Bertram G. Murray, Jr./Animals Animals; A69 (l) Gilbert S. Grant/Photo Researchers; A69 (r) J.A. Kraulis/Masterfile; A70 Christi Carter/Grant Heilman Photography; A72 H. Mark Weidman; A73 Porterfield-Chickering/Photo Researchers; A73 (i) Dr. Jeremy Burgess/Science Photo Library/Photo Researchers; A74 (t) Frans Lanting/Minden Pictures; A74 (b) Patti Murray/Earth Scenes; A75 C.K. Lorenz/Photo Researchers; A76 Dr. E.R. Degginger/Color-Pic; A78 (tl), A78(bl) Runk/Schoenberger/Grant Heilman Photography; A78 (br) E.R. Degginger/Earth Scenes; A79 (t), A79 (ti) Runk/Schoenberger/Grant Heilman Photography; A79 (b) Jessie M. Harris; A79 (bi) Steve Solum/Bruce Coleman, Inc.; A80 (r) Bill Lea/Dembinsky Photo Associates; A80 (tl), A80 (bl) Kim Taylor/Bruce Coleman, Inc.; A82 Runk/Schoenberger/Grant Heilman Photography; A84 Gregory K. Scott/Photo Researchers; A86 Angel/Biofotos; A86 (l) Runk/Schoenberger/Grant Heilman Photography; A88 James Lyle/Texas A&M University; A89 Chris Rogers/The Stock Market; A90 Hunt Institute for Botanical Documentation/Carnegie Mellon University; A93 (t) Bruce Coleman, Inc.; A93 (b) Breck P. Kent/Animals Animals; A93 (b) Denise Tackett/Tom Stack & Associates; A96 (tr) Al Lamme/Len/Phototake; A96 (br) Astrid & Hanns-Frieder Michler/Science Photo Library/Photo Researchers; A96 (cr) M. Abbey/Photo Researchers; A98 Biophoto Associates/Science Source/Photo Researchers; A100 (tr) Al Lamme/Len/Phototake; A100 (br) Astrid & Hanns-Frieder Michler/Science Photo Library/Photo Researchers; A100 (cr) M. Abbey/Photo Researchers; A110 Biophoto Associates/Photo Researchers; A114 Produced with permission from ETHICON, INC., 2001 Somerville, NJ; A115 Owen Franken/Stone; A116 (t) UPI/Corbis; A116 (b) Will and Deni McIntyre/Photo Researchers; A120 (t) Keren Su/Corbis; A120 (b) Jan Butchofsky/Corbis.

Unit B

Unit B Opener (fg) Steve Kaufman/DRK Photos; (bg) H. Richard Johnson/FPG International; B2-B3 Kim Heacox/Peter Arnold, Inc.; B3 (l) G. Perkins/Visuals Unlimited; B3 (r) Herbert Schwind/Okapia/Photo Researchers; B4 Navaswan/FPG International; B6-B7 D. Logan/H. Armstrong Roberts, Inc.; B8 (t) Mark E. Gibson/Dembinsky Photo Associates; B8 (b) Larry Lefever/Grant Heilman Photography; B9 (t) Mark E. Gibson/Dembinsky Photo Associates; B10 Dr. E. R. Degginger/Color-Pic; B12 (bg) Doug Cheeseman/Peter Arnold, Inc.; B12 (li) Dr. E. R. Degginger/Color-Pic; B12 (ri) FarrellGrehan/Photo Researchers; B13 (bg) Luiz C. Marigo/Peter Arnold, Inc.; B13 (ri) M. Timothy O'Keefe/Bruce Coleman, Inc.; B13 (li) Bernard Boutrit/Woodfin Camp & Associates; B14 (bg) Jim Steinberg/Photo Researchers; B15 (t) M. Timothy O'Keefe/Bruce Coleman, Inc.; B16 (t) Tom Bean/The Stock Market; B16 (b) Grant Heilman Photography; B18 Ken M. Highfill/Photo Researchers; B20 (bg) Rob Lewine/The Stock Market; B20 (ci) Ken Brate/Photo Researchers; B20 (li) Art Wolfe/Stone; B20 (ri) Matt Meadows/Peter Arnold, Inc.; B21 (tl) David M. Phillips/Photo Researchers; B21 (tr) R & J Spurr/Bruce Coleman, Inc.; B21 (bl) Dwight R. Kuhn; B21 (br) Institut Pasteur/Phototake; B21 (cl) E. R. Degginger/Photo Researchers; B21 (cr) Lewis Kemper/Stone; B22-B23 (bg) Carr Clifton/Minden Pictures; B24 (t) David Carriere/Stone; B24 (b) E & P Bauer/Bruce Coleman, Inc.; B25 Roy Morsch/The Stock Market; B26 Stuart Westmorland/Stone; B29 (t) James Martin/Stone; B29 (c) Dr. E. R. Degginger/Color-Pic; B29 (b) Tom McHugh/Photo Researchers; B31 (t) Manfred Kage/Peter Arnold, Inc.; B31 (c) William Townsend, Jr./Photo Researchers; B31 (b) Norbert Wu/The Stock Market; B32 Dr. E. R. Degginger/Color-Pic; B33 Tom McHugh/Photo Researchers; B34 Francois Gohier; B36-B37 (bg) Jack McConnell; B37 (bli) Susie Leavines/Picture It!; B37 (bri) Jeff Foott/Bruce Coleman, Inc.; B37(tli) Gay Bumgarner; B37(tri) Dr. Charles Steinmetz, Jr.; B37 (ci) Gregory G. Dimijian/Photo Researchers; B38 (b) Fred Bavendam/Minden Pictures; B38 (t) Flip Nicklin/Minden Pictures; B39 (b) Dr. E. R. Degginger/Color-Pic; B39 (t) Flip Nicklin/Minden Pictures; B40 (b) University of Delaware Graduate School of Marine Studies; B42 Oakridge National Laboratory; B43 Norbert Wu; B43 (b) David Young-Wolff/PhotoEdit; B44 (b) David Muench/Stone; B44 (t) Courtesy of Indiana Dunes National Lakeshore/National Park Service; B48 Joel Sartore from Grant Heilman Photography; B48-B49 Jeff Foott/Bruce Coleman, Inc.; B49 (b) Roy Toft/Tom Stack & Associates; B50 Gary Braasch/Woodfin Camp & Associates; B53 MikeYamashita/Woodfin Camp & Associates; B54 (t) Frank Oberle/Stone; B54 (bl) David Woods/The Stock Market; B54 (bri) Stan Osolinski/Dembinsky Photo Associates; B54-B55 (b) Jeff Henry/Peter Arnold, Inc.; B55 (t) Milton Rand/Tom Stack & Associates; B55 (ti) Joe McDonald/Tom Stack & Associates; B55 (bi) Stan Osolinski/Dembinsky Photo Associates; B56 (t) Aneal Vohra/Unicorn Stock Photos; B56 (b) Merrilee Thomas/Tom Stack & Associates; B56 (bi) Tom Benoit/Stone; B57 Mark E. Gibson; B58 Tom Walker/Stock, Boston; B60-B61 (t) David Harp Photography; B60-B61 (b) Jason Hawkes/Stone; B61 (tr) Frans Lanting/Stone; B61 (br) Scott Slobodian/Stone; B62 (t) Mark E. Gibson;B62-B63 (b) Paul Chesley/Stone; B63 (t) J. Lotter/Tom Stack & Associates; B64 (t) David R. Frazier; B64 (b) Lori Adamski Peek/Stone; B65 Marc Epstein/Visuals Unlimited; B66 Jim Schwabel/New England Stock Photo; B69 (b) SuperStock; B70 (c) A. Bolesta/H. Armstrong Roberts, Inc.; B70 (r) IPP/H. Armstrong Roberts, Inc.; B71 (b) Mark Joseph/Stone; B72 (t) Gary Braasch Photography; B72 (b) Barbara Gerlach/Dembinsky Photo Associates; B73 Superstock; B74 (t) J. Blank/H. Armstrong Roberts, Inc.; B74 (bl) R. Kord/H. Armstrong Roberts, Inc.; B74 (br) Marc Muench/David Muench Photography; B75 (b) G. Ahrens/H. Armstrong Roberts, Inc.; B76 AP/Wide World Photos; B78 Earl Roberge/Photo Researchers; B80 (t) Jon Gnass; B80 (b) Connie Toops.

Unit C

Unit C Opener (fg) Francois/Gohier/Photo Researchers; C3 (b) Packwood, R./Earth Scenes; C4 Jock Montgomery/Bruce Coleman, Inc.; C8 Kenneth Fink/Bruce Coleman, Inc.; C9 Eric A. Wessman/The Viesti Collecton; C10-C-11 Kevin Schafer; C12 Wesley Bocxe/Photo Researchers; C14 Francois Gohier/Photo Researchers; C16 (t) Russell D. Curtis/Photo Researchers; C16 (bl) Tom McHugh/Photo Researchers; C16 (bc) Lee Foster/Bruce Coleman, Inc.; C17 Francois Gohier/Photo Researchers; C18 E.R. Degginger/Photo Researchers; C22-C23 C. C. Lockwood/Bruce Coleman, Inc.; C23 (t) Paolo Koch/Photo Researchers; C24 (t) Photo Researchers; C24 (b) H. Hannejdottir/Photo Researchers; C24 (bg) Ken Sakamoto/Black Star; C25 Krafft-Explorer/Photo Researchers; C26 Bill Ingals/NASA/Sygma; C27 Nancy Simmerman/Stone; C28 (t) Todd Bigelow/Black Star; C28 (b)Thomas Jaggar/NGS Image Collection; C32-C33 Gerd Ludwig/Woodfin Camp & Associates; C33 (t) The Westthalian Museum of Natural History; C33 (b) James Martin/Stone; C34 Murray Alcosser/The Image Bank; C36 (t) Beth Davidow/Visuals Unlimited; C36 (b) The Natural History Museum, London; C36-C37 (b) William E. Ferguson; C37 (c) The Natural History Museum, London; C37 (tl) Dr. P. Evans/Bruce Coleman Collection; C37 (tr) David J. Sams/Stone; C37 (l), C38 (l) William E. Ferguson; C38 (t) Peter Gregg/Color-Pic; C38-C39 (t) Bob Burch/Bruce Coleman, Inc.; C38-C39 (b) E. R. Degginger/Bruce Coleman, Inc.; C39 (ti) E. R.Degginger/Bruce Coleman, Inc.; C40, C42 (l) The Natural History Museum, London; C42 (li) William E. Ferguson ; C42-C43 (bg) Mark J.Thomas/Dembinsky Photo Associates; C43 (t) Runk/Schoenberger/Grant Heilman Photography; C44 (t) Phil Schofield/Stone; C44 (c) Phil Degginger/Bruce Coleman, Inc.; C45 Louie Psihoyos/Matrix International; C46 (t) Kevin Fleming; C46 (br) Breck B. Kent; C46 (b) David Hiser/PictureQuest; C47 (t), C47 (i) Layne Kennedy; C47 (b) Richard T. Nowitz/Corbis; C48 (t) Ted Clutter/Photo Researchers; C48 (b) J. C. Carton/Bruce Coleman, Inc.; C49 William E. Ferguson; C50 J.C. Carton/Bruce Coleman, Inc.; C52 (l) Larry Lefever from Grant Heilman Photography; C52 (r) Thomas Kitchin/Tom Stack & Associates; C53 (t) Kenneth Murray/Photo Researchers; C55 (t), C55 (b) Dr. E.R. Degginger/Color-Pic; C55 (tc), C55 (bc) Breck P. Kent/Earth Scenes; C58 (b) The Granger Collection, New York; C58 (tl) Alinari/Art Resource, NY; C58 (tr) The Granger Collection, New York; C58 (t) Bill Bachman/Photo Researchers; C59 (b) James King-Holmes/Science Photo Library/Photo Researchers; C60 (t) San Francisco State University/Department of Geosciences; C60 (b) AP/Wide World Photos; C64 (t)

J.F. Maxwell/Falls of the Ohio State Park; C64 (b) Sandy Felsenthal/Corbis.

Unit D

Unit D Opener (fg) Dorling Kindersley; (bg) FPG International; D2-D3 Warren Faidley/International Stock; D3 (t) Bob Abraham/The Stock Market; D3 (b) NRSC Ltd/Science Photo Library/Photo Researchers; D4 Keren Su/Stock, Boston; D6 Space Frontiers-TCL/Masterfile; D10 Bruce Watkins/Earth Scenes; D12 Peter Menzel/Stock, Boston; D14-D15 C. O'Rear/Corbis; D15 Warren Faidley/International Stock; D16 (b) Bill Binzen/The Stock Market; D18 J. Taposchaner/FPG International; D20 Sam Ogden/Science Photo Library/Photo Researchers; D21 (t) Breck P. Kent/Earth Scenes; D21 (b) B. Daemmrich/The Image Works; D22 (c) 1998 Accu Weather; D24 Geophysical Institute, University of Alaska, Fairbanks/NASA; D25 Pat Lanza/Bruce Coleman, Inc.; D26 (t) Clark Atlanta University; D26 (b) NASA/Science Photo Library/Photo Researchers; D30-D31 Warren Bolster/Stone; D31 (t) Warren Morgan/Corbis; D31 (b) Tom Van Sant, Geosphere Project/Planetary Visions/Science Photo Library/Photo Researchers; D32 Philip A. Savoie/Bruce Coleman, Inc.; D36 (tr) A. Ramey/Stock Boston; D36 (bl) Richard Gaul/FPG International; D38 John Lel/Stock, Boston; D40 E.R. Degginger/Photo Researchers; D41 (t) Fredrik Bodin/Stock, Boston; D41 (b) Peter Miller/Photo Researchers; D42 (t), D42 (c) Francois Gohier/Photo Researchers; D42 (b) Steinhart Aquarium/Tom McHugh/Photo Researchers; D46 Ralph White/Corbis; D49 (i) Dr. E.R. Degginger/Color-Pic; D49 (t) David R. Frazier; D50-D51 Marie Tharp/Oceanic Cartographer; D54-D55 Ben Margot/AP Photo/Wide World Photos; D55 Thomas Ives/The Stock Market; D56 (t) Erich Hartmann/Magnum Photos; D56 (b) Ron Sefton/Bruce Coleman, Inc.; D60-D61 Ton Kinsbergen/ESA/Science Photo Library/Photo Researchers; D61 NASA; D62 Dr. E. R. Degginger/Color-Pic; D68 Tom Till; D71 Frank Zullo/Photo Researchers; D72-D73 M. Agliolo/Photo Researchers; D74 NASA; D77 (t) U.S. Geological Survey/Science Photo Library/Photo Researchers; D77 (b) David Crisp and the WFPC2 Science Team (Jet Propulsion Laboratory/California Institute of Technology)/NASA; D77 (tc) NASA; D77 (bc) National Oceanic and Atmospheric Administration; D78 NASA; D78-D79 Erich Karkoschka (University of Arizona Lunar & Planetary Lab) and NASA; D79 (r) Dr. R. Albrecht, ESA/ESO Space Telescope European Coordinating Facility, NASA; D79 (c) Lawrence Sromovsky (University of Wisconsin - Madison), NASA; D80, D81, D82 NASA; D84 (r) Michael Freeman; D84 (tl) David Nunuk/Science Photo Library/Photo Researchers; D84 (bl) Omikron Collection/Photo Researchers; D85 (t) Simon Fraser/Science Photo Library/Photo Researchers; D85 (b) Robert Frerck/Stone; D85 (ti) Roger Ressmeyer/Corbis; D86 (bg) Shahn Kermani/Liaison International; D86, D87, D88, D89 NASA; D90 (r) Jean-Loup Charmet; D90 (l) NASA; D91 (t) The Granger Collection, New York; D91 (c) Sylvester Allred/Visuals Unlimited; D91 (b) Mark E. Gibson/Dembinsky Photo Associates; D92 J. Kelly Beatty; D92 (bg) Science VU/Visuals Unlimited; D96 (t) Mark E. Gibson; D96 (b) W. Metzen/H. Armstrong Roberts, Inc..

Unit E

Unit E Opener (fg); Dennis Yankus/Superstock; (bg); Pierre-Yves Goavec/Image Bank; E2-E3 Jon Riley/Stone; E3 Dr. E.R. Degginger/Color-Pic; E4 Superstock; E6 Michael Denora/Liaison International; E8 (b) Bob Abraham/The Stock Market; E10 (l) Lee F. Snyder/Photo Researchers; E14-E15 (b) Richard R. Hansen/Photo Researchers; E16 Stone; E20 (t) Kathy Ferguson/PhotoEdit; E20 (b) Doug Perrine/Innerspace Visions; E20 (bi) Felicia Martinez/PhotoEdit; E21 (b) Chip Clark; E22 (bg) Norbert Wu/Mo Yung Productions; E22-E23 Richard Pasley/Stock, Boston; E24 Stockman/International Stock; E29 (cl) Dr. E.R. Degginger/Color-Pic; E29 (br) Grace Davies; E29 (bl) Index Stock Imagery/PictureQuest; E30 (i) P. Degginger/H. Armstrong Roberts, Inc.; E30 (b) Jack McConnell/McConnell & McNamara; E30-E31 Paul A. Souders/Corbis; E32 Courtesy of J. G.'s Edible Plastic; E33 David R. Frazier; E34 (l) United States Nuclear Regulatory Commission; E34 (r) Tom Carroll/Phototake; E38-E39 Ray Ellis/Photo Researchers; E39 (t) Peter Steiner/The Stock Market; E39 (b) Murray & Assoc./The Stock Market; E40 Craig Tuttle/The Stock Market; E43 (t) Jim Zipp/Photo Researchers; E44 Ted Horowitz/The Stock Market; E48 D. Nabokov/Gamma Liaison; E50 (b) L. West/Bruce Coleman, Inc.; E50 (i) Jonathan Wright/Bruce Coleman, Inc.; E51 Gary Milburn/Tom Stack & Associates; E52 Jeff Foott/Bruce Coleman, Inc.; E56 (t) Craig Hammell/The Stock Market; E56 (b) Russell D. Curtis/Photo Researchers; E57 (tl) Stu Rosner/Stock, Boston; E57 (tr) John Mead/Science Photo Library/Photo Researchers; E57 (br) John Cancalosi/Stock, Boston; E58 (b) David Falconer& Associates; E58 (tr) Telegraph Colour Library/FPG International; E58 (cr) Charles D. Winters/Photo Researchers; E58 (bi) Montes De Oca & Associates; E59 Paul Shambroom/Science Source/Photo Researchers; E61 Danny Daniels/The Picture Cube; E62 Minnesota Historical Society; E62 (i) Peter Vadnai/The Stock Market; E66-E67 Stephen Dalton/Photo Researchers; E67 Carl R. Sams, II/Peter Arnold, Inc.; E68 A. Ramey/PhotoEdit; E71 (b) Summer Productions; E71 (li) Michelle Bridwell/PhotoEdit; E71 (tri) Peter Langone/International Stock; E72 (l) Ian O'Leary/Stone; E76 Randy Duchaine/The Stock Market; E78 (r) Jim Zipp/Photo Researchers; E78 Spencer Grant/PhotoEdit; E82 (t) Bose Corporation; E82 (b) Bose/Lisa Borman Associates; E84-E85 (b) NASA; E90-E91 Bruce Forster/Stone; E91 (i) Jon Riley/Folio; E96-E97 NASA; E97 (t) Chip Simons; E97 (b) David Madison/Bruce Coleman, Inc.; E100 (l) Mark E. Gibson; E100 (r) Bob Daemmrich/Stock, Boston; E103 Jan Butchofsky/Dave G. Houser; E105 (t), 105 (c) Richard Megna/Fundamental Photographs; E108 (b) Randy Duchaine/The Stock Market; E110 Tom Skrivan/The Stock Market; E111 (tr) David Woodfall/Stone; E113 Roy Morsh/The Stock Market; E114-F115 Paul Silverman/Fundamental Photographs; E115 (t) Ed Eckstein for the Franklin Institute Science Museum; E115 (b) Peter Angelo Simon/The Stock Market; E116 (t) Schomburg Collection; E116 (bl) Jim Davie; E120 (t) Sal Dimarco/Black Star; E120 (b) Jack Olson.

Unit F

Unit F Opener (fg) Steve Berman/Liaison International; (bg) Photone Disk #50 ; F2-F3 Pete Saloutos/The Stock Market; F4 Doug Martin/Photo Researchers; F9 Charles D. Winters/Photo Researchers; F10 Cosmo Condina/Stone; F15 Dr. E.R. Degginger/Color-Pic; F16 National Maritime Museum Picture Library; F19 Richard Megna/Fundamental Photographs; F20-F21 (t) Phil Degginger/Color-Pic; F22 Gamma Tokyo/Liaison International; F24-F25 Spencer Grant/PhotoEdit; F25 Tom Pantages; F27 (t), F27 (c) Phil Degginger/Color Pic; F27 (b) Bruno Joachin/Liaison International; F27 (bg) W. Cody/Corbis Westlight; F30 (t) PhotoDisc; F30 (b) Corbis-Bettmann; F31 Phil Degginger/Color-Pic; F32 (i) Fonar Corporation; F32 Jean Miele/The Stock Market; F36-F37 PictureQuest; F37 (t), F37 (ti) Dwight R. Kuhn; F37 (br) Tony Freeman/PhotoEdit; F38 (b) David R. Frazier; F40 (b) Mark E. Gibson; F42 (b) Bob Daemmrich/Stock, Boston; F44 (bl) Miro Vintoniv/Stock, Boston; F46 (bl) Jean-Marc Barey/Agence Vandystadt/Photo Researchers; F47 (tr) Daniel MacDonald/The Stock Shop; F49 Bernard Asset/Agence Vandystadt/Photo Research; F50 (t) Kathi Lamm/Stone; F53 Terry Wild Studio; F54 (b) William R. Sallaz/Duomo Photography; F57 (c) Photo Library International/ESA/Photo Researchers; F57 (cr) Photo Researchers; F58 (bl) Michael Mauney/Stone; F60 (b) Brian Wilson; F60 (t) PA News; F61 (b) Michael Newman/PhotoEdit; F62 (t) UPI/Corbis-Bettmann; F62 NASA; F66-F67 (bg) Dan Porges/Bruce Coleman, Inc.; F67 (tr) R. Sheridan/Ancient Art and Architecture Collection; F67 (tl) The Granger Collection, New York; F68 (bl) William McCoy/Rainbow; F70 (bl) Yoav Levy/Phototake/PictureQuest; F73 (tr) Michael Newman/PhotoEdit; F74 (tr) David R. Frazier; F78 (bl) Mark E. Gibson; F80 (b) Jeff Dunn/Stock, Boston; F82 (bl) Tom King/Tom King, Inc.; F84 (b) Aaron Haupt/David R. Frazier; F85 (tl) Dan McCoy/Rainbow; F85 (br) Michael Newman/PhotoEdit; F85 (cl) David Falconer/Folio; F86 (t) Superstock; F87 (r) Churchill & Klehr; F87 (bl) Staircase & Millwork Corporation, Alpharetta, GA; F88 (br) Tony Freeman/PhotoEdit; F90 (bg) Corel; F90 (l) Archive Photos; F90 (c) Noble Stock/International Stock; F90 (bl) Alexandra Guest/John F. Coates; F91 (l) Eric Sanford/International Stock; F94 (l) Archive Photos; F92 (r) Library of Congress/FPG International; F92 (bl) Library of Congress; F96 (t) Christian Heeb/Gnass Photo Images; F96 (b) Maxine Cass.

Health Handbook: R23 Palm Beach Post; R27 (t) Andrew Spielman/Phototake; (c) Martha McBride/Unicorn Stock; (b) Larry West/FPG International; R28 (l) Ron Chapple/FPG; (c) Mark Scott/FPG; (b) David Lissy/Index Stock.

All other photographs by Harcourt photographers listed below, © Harcourt: Weronica Ankarorn, Bartlett Digital, Victoria Bowen, Eric Camden, Digital Imaging Group, Charles Hodges, Ken Karp, Ken Kinzie, Ed McDonald, Sheri O'Neal, Terry Sinclair.

Art Credits

Mike Atkinson A85, B22, B23; Jean Calder A99, A100 - 101, A110, A111; Susan Carlson D22; Mike Dammer A33, A65, A91, A117, B45, B77, C29, C61, D27, D57, D93, E35, E63, E93, E117, F33, F63, F93; John Edwards E111; John Francis B14 - 15; Lisa Frasier E56-57; George Fryer C6 - 7, C20, C21, D20; Thomas Gagliano D48 - 49, E73, E78, E79, E80, F14; Patrick Gnan E60, F74; Terry Hadler E14, E19, E49, F78, F79; Tim Hayward C44, C48; Robert Hynes A16, A22, B28 - 29, B30 - 31; Joe LeMoniier A64, A90, E66; Mapquest A57, B40, C46; Janos Marffy D66; Michael Maydak B38 - 39; Sebastian Quigley D12 - 13, D44, D65, D76 - 77, D78 - 79, E6, E7, E8, E22, E72, F6, F7, F20, F24, F56; Eberhard Reinmann A98, A104 - 105, A106, A112, E74; Mike Saunders A7, A8, A20, A26, A27, A73, A84, B52 - 53, C7, C36, D8 - 9, D52 - 53, D70, D71, D72 - 73; Steve Seymour B7, B40 - 41, B62, B63, C8, C9, C10, C15, D14, D43, D86, E43, E44, E51, E52, E70, E85, E88 - 89, F8, F13, F49, F72; Shough E112; Bill Smith Studio D93; Walter Stuart A10 - 11, A14, A15; Steve Weston C14, C22, C23, D15, E72, E86 - 87, F18, F80, F86, F87, F88, F93